Dissertation Research and
Writing for Construction Students

D0161033

To my mother, Rose

Dissertation Research and Writing for Construction Students

Second edition
Dr S. G. Naoum

AMSTERDAM · BOSTON · HEIDELBERG · LONDON · NEW YORK · OXFORD
PARIS · SAN DIEGO · SAN FRANCISCO · SINGAPORE · SYDNEY · TOKYO

Butterworth-Heinemann is an imprint of Elsevier

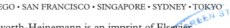

ELSEVIER

Butterworth-Heinemann is an imprint of Elsevier
Linacre House, Jordan Hill, Oxford OX2 8DP, UK
The Boulevard, Langford Lane, Kidlington, Oxford OX5 1GB, UK
84 Theobald's Road, London WC1X 8RR, UK
Radarweg 29, PO Box 211, 1000 AE Amsterdam, The Netherlands
30 Corporate Drive, Suite 400, Burlington, MA 01803, USA
525 B Street, Suite 1900, San Diego, CA 92101-4495, USA

First edition 1998
Reprinted 1998, 1999 (twice), 2001 (twice), 2002, 2003 (twice), 2006
Second edition 2007

British Library Cataloguing in Publication Data
Naoum, S. G. (Shamil G.)
    Dissertation research and writing for construction
    students. – 2nd ed.
    1.   Dissertations, Academic 2.   Technical writing
    I.   Title
    808'.066624

Library of Congress Control Number: 2006 929340

ISBN–13: 978-0-7506-8264-0
ISBN–10: 0-7506-8264-7

For information on all Butterworth-Heinemann publications visit our
website at www.books.elsevier.com

Typeset by Charon Tec Ltd, (A Macmillan Company), Chennai, India
www.charontec.com

Printed and bound in the UK

07 08 09 10 11 11 10 9 8 7 6 5 4 3 2 1

Contents

Preface

This book has grown out of my involvement in supervising BSc and MSc dissertations as well as my accumulated experience of teaching research methods to postgraduate students at London South Bank University. Over the years, I became convinced that construction students needed something akin to research methods lectures, in addition to tutoring, that could provide them with guidance on basic research techniques and how to write a dissertation. There have been an enormous number of text books which presented research design and methodologies but few, if any, are related to built environment studies.

This book approaches dissertation research and writing as a process, involving a sequence of learnable activities. Each activity prepares the student for the next step and some steps are best taken before others. I have divided the process of dissertation research into three parts, which are best carried out in the order presented.

Part one is 'preparing the ground'. It involves:

1 Identifying the problem (including narrowing and clarifying the problem);
2 Writing a proposal (including the aim, objectives, hypothesis and/or key research questions;
3 Reviewing the literature (including critical appraisal of literature).

Part two is 'research design'. It involves:

4 Approaches to data collection (deciding whether to use a survey or a case study);
5 Techniques for data collection (deciding whether to use the post or go for interviews);
6 Constructing and sampling the questionnaire.

Part three is 'analysis of the results' and producing the dissertation. It involves:

7 Measurement of data;
8 Analysis of data (including interpretation and discussion of results).
9 Structuring and writing the whole dissertation;
10 Dissertation supervision and assessment.

This book is aimed specifically at BSc and MSc students, who are embarking on research as part of their degree. It will provide students with a clear explanatory text which is supported by numerous examples illustrating good practice. Students of Built Environment subjects such as Construction Management, Facility Management, Building Surveying, Quantity Surveying and Civil Engineering will find this book of use. It will also be of use to those candidates requiring membership of the Chartered Institute of Building (CIOB) and anyone else involved in research work.

About the second edition

It has been very pleasing to receive the many complimentary and positive comments from reviewers and users of the first edition. These comments have been most appreciated and provided the encouragement for, the basis of, the second edition.

The second edition retains the same underlying theme, aims and approaches. It also carriers forward the same basis framework with the same sequencing and ordering of the chapters.

There is however a general review and update of material. The rationalisation of the text recognises the need for a new chapter on dissertation supervision and assessment.

About the author

Dr Shamil Naoum is a senior lecturer at London South Bank University. He received a bachelor of science in Building and Construction Engineering from the University of Technology in Baghdad, a master of science in Construction Management and Economics from the University of Aston in Birmingham, and a PhD in Construction Management from Brunel University in Middlesex.

Before beginning his academic career he worked in the construction industry as a site engineer and project manager. He is a full member of the Chartered Institute of Building and the American Society of Civil Engineers.

Dr Naoum has considerable research experience in construction management related areas, such as procurement methods, human resources management and management science. In addition to supervising BSc and MSc dissertations, Dr Naoum works with PhD students researching construction management problems.

During his academic career, Dr Naoum has published papers in many international conferences and a diverse group of scholarly journals, including: *American Society of Civil Engineers* (USA); *International Journal of Project Management* (UK); *Journal of Engineering Construction and Architectural Management* (UK) and *Chartered Institute of Building* (UK).

Acknowledgements

I wish to express my special thanks and appreciation to David Coles for his encouragement during the preparation of this book.

I am also grateful to the following students for their permission to extract information from their dissertations: Miss H. Cooper; Mr S. Green; Mr M. Howard; Miss D. Manning and Mr. F. Rassam.

I also wish to thank Longman Group Ltd, London, for permission to adapt appendices from their book *Statistical Tables for Biological, Agricultural, and Medical Research*, sixth edition (1974); to Pitman Publishing, for permission to adapt a table from their book, *Management and Organisational Behaviour*, fourth edition (1996) and to Unwin Hyman Publications, for permission to adapt a table from their book, *Quantity and Quality in Social Research* (1988).

Finally, I wish to express my gratitude to the staff of Butterworth-Heinemann who masterfully crafted the production of this book.

Dr Shamil Naoum
Department of Surveying, Property and Construction
London South Bank University

1 Introduction

This book provides a guide and learning support document in the preparation of a dissertation for honours undergraduate students and for students undertaking dissertations on taught Masters degrees. The book focuses specifically on built environment related subjects with a special reference to the discipline areas of Construction, Project Management and Building. However, as the book follows the interrelated process of conducting a research, it can also be useful to undergraduate and postgraduate students of various disciplines. It covers issues such as the selection of a dissertation topic, writing a proposal, conducting a literature review, selecting the research approach, devising research instruments, collecting information, analysing and presenting information and producing a well-written dissertation.

Rationale for conducting a dissertation

The rationale for including a dissertation as a major component of the BSc and MSc courses is that it provides for the development of intellectual skills of a kind that are not fully facilitated on the other components of the teaching course. The dissertation requirement accords with the educational philosophy of the BSc and MSc course in that it requires students to take responsibility for their own learning, specifying and defining the task, and defining the learning outcomes.

The aims and objectives for conducting a dissertation are therefore:

1 To provide the opportunity for students to undertake an independent piece of work of a demanding nature.
2 To enable students to investigate, in depth, a subject in which they have a particular interest and of their own selection.
3 To provide students with an opportunity to stretch their intellectual and technical skills.

4 To encourage the students to develop new forms of analysis, conclusions and policies which may make an original contribution to the knowledge in the field of study concerned.
5 To encourage both clarity and depth of thought in that the project involves analysis of a problem in depth and the development of a logical sequence of ideas.
6 To provide students with an opportunity of learning how to acquire detailed information on a particular issue. It will involve them in the use of bibliographies, libraries and library reference systems. It will involve them in using primary sources of data such as a census, and it will involve the collection of new data through interviews, surveys and archival research.
7 To require the proper use, presentation and communication of data.

The meaning of research

The *Concise Oxford Dictionary* defines research as 'careful search or inquiry; endeavour to discover new or collate old facts etc. by scientific study of a subject; course of critical investigation'. From which we can conclude that the word 'research' may be used interchangeably with 'inquiry', 'study' or 'investigation'. And yet something more is implied: the inquiry, study or investigation must be conducted in a careful, scientific and/or critical manner. Others have added to this list of adjectives methodical and systematic. It does not really matter which combination of words is chosen (systematic inquiry or critical investigation) so long as both words are kept clearly in mind.

In addition to a specific method of inquiry, a research project has to have an aim or objective. In general terms, the aim of all research is to expand knowledge. But we do not simply want a list of facts. A good piece of research will focus on certain aspects of a topic. It will seek to answer specific questions, solve a particular problem or test a hypothesis. The issue(s) to be addressed must be clearly stated at the outset in the objective(s) of research.

Purpose of the book

This book fulfils the need of the construction industry by providing students with a useful guide to undertake a piece of research. The construction industry is now developing a community with a new attitude to research. Although a number of texts are available on research design and methodology, there is little, if any, applied in construction. This book is specifically designed to assist:

1 Honours undergraduate students in the Built Environment with specific emphasis on Construction related degree programmes.
2 CIOB (member) new Educational framework course programmes offered by universities/colleges of higher educational/correspondence courses, etc.
3 CIOB (direct member) thesis route programme.
4 MSc (taught Masters) students in Construction Management/Project Management related subjects.

The book will also assist:

1 Students of other related Built Environment disciplines, e.g. Building Surveying, Quantity Surveying, Architecture, Building Economics, etc.
2 Students in the social sciences and people concerned with social surveys.
3 Additionally, the book should provide a useful 'foundation' guide to students who are about to start an MPhil/PhD programme. However, it is not designed to provide a manual for PhD students nor does it provide the type of analysis that a doctorate degree deserves.

The plan of the book

As mentioned earlier, this book will follow the interrelated stages of conducting a dissertation research. Figure 1.1 illustrates the research process stages with their corresponding chapters. This book is therefore composed of ten chapters.

Chapter 1: Introduction

This chapter has been a general introduction to dissertation research and outlines the main aims and objectives of the book.

Chapter 2: Selecting a topic and writing the dissertation proposal

The choice of a dissertation topic usually comes from your interest in and value of a particular subject, which are usually interrelated. The interest and value

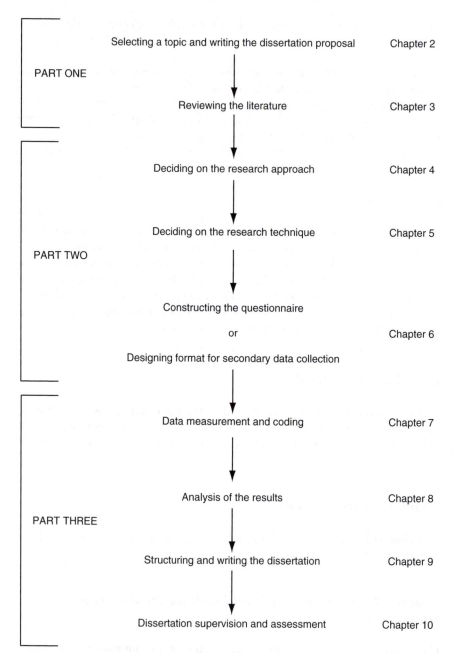

Figure 1.1   *Research process diagram*

will eventually be developed into a series of questions which you will be keen to find answers to. Selecting a topic is discussed in the first part of the chapter. After selecting the subject of your dissertation, you need to formulate a proposal which should contain a rationale, aim, objectives, hypothesis (or key questions), methodology and a programme of work. The extent and degree of details for the dissertation proposal are given in the second part of this chapter.

Chapter 3: Reviewing the literature

This activity will most likely be carried out throughout the whole research process but more extensively at the earlier stages of the research. It basically involves reading and critically appraising what other people have written about your subject area. The chapter also gives details on how to conduct a systematic literature search.

Chapter 4: Approaches to data collection

After deciding on the topic that you wish to study and having conducted an extensive literature search, you will be in a position to design your research. Designing the research involves the following activities:

1   Deciding on the type of data that has to be collected (quantitative or qualitative data). This chapter provides further description of these terms.
2   Confirm the method of data collection, i.e. should you conduct a survey or a case study. The chapter describes each of these methods in great detail.
3   Deciding on the techniques for data collection, i.e. should you gather the data by interviews, by telephone or by postal questionnaire.
4   Designing your sample. The research design should tell you how your sample needs to be drawn, to whom you should target your questionnaire and how many to issue.

Chapter 5: Techniques for data collection

After deciding on your research approach you will be involved in collecting the data for your research study. If you are collecting 'primary' data (data collected at first hand), you need to undertake certain activities such as inviting your sample to complete the questionnaire, preparing the data summary sheet, or to take whatever action is required for your fieldwork. If you are collecting 'secondary' data (data gathered from some other sources), you need to contact those organisations who store the data (if not stored in the library) such as the state and its agencies, statistical offices like Her Majesty's Stationery Office (HMSO) and the Health and Safety Executive (HSE). Chapter 5 explains the main features of the postal survey and the interview technique together with showing the advantages and limitations of both techniques.

Chapter 6: Questionnaire construction

This chapter describes and illustrates methods of questionnaire construction: the content of questions, types of questions, question format and the sequence of questions.

Note to students: After constructing your questionnaire, it is strongly recommended to conduct a trial run on the questionnaire before circulating it to the whole sample. This trial is called a 'pilot study'. The pilot study involves testing the wording of the questions, establishing the length of the questionnaire, avoiding ambiguous questions, suggestions for analysing the data, as well as testing the technique selected for collecting the data. In large research projects such as MPhil or PhD, the researcher should try to even analyse the results of the pilot study to ensure the reliability of data collected. Chapter 6 gives further details on the pilot study.

Chapter 7: Measurements and probability

This chapter prepares the ground for analysis of your results. It is concerned with the nature of measurement in research. It includes the explanation of the four levels of measurement: nominal, ordinal, interval and ratio. The main point to recognise about measurement is that each level of measurement requires a certain type of analysis that is more appropriate than others. This chapter will also explain the meaning of the term 'probability' and its relevance to the analysis of your results.

Chapter 8: Analysis of the results

Once you have collected all the completed questionnaires and/or gathered the archival data, you will be ready for the next stage – the analysis of the data and determining the direction of the study. This stage involves processing the data, putting answers to categories and generally finding out the pattern of the responses. Some results require statistical analysis, as in the case of a large sample survey, and some results involve finding out the trend of the responses, as in the case of in-depth interviews. Chapter 8 gives details on how to analyse the results.

Chapter 9: Structuring and writing the dissertation

After you have completed the literature review and analysed the data, you will be ready to write up the whole dissertation. Writing your dissertation involves reporting and critically appraising the literature review, analysis of the data, discussion and the interpretation of your findings. Bear in mind that the writing up stage can be overlapped with any of the above activities and may start as early as the literature review stage. Chapter 9 provides details on structuring and writing a typical dissertation project.

Chapter 10: Dissertation supervision and assessment

Once you have decided on the topic of your study and your proposal gets approved, your department should then appoint a personal supervisor for you. It is therefore important to know the role of your supervisor and what is expected from him/her.

This chapter will discuss the basics of what you should expect from your supervisor. It first explains what the role of the supervisor is and what it is not. Second, it illustrates how you may plan your dissertation and finally how you would expect to be assessed in the end.

Part One

Preparing the Ground and Reviewing the Literature

2 Selecting a topic and writing the dissertation proposal

The first step of the research process involves selecting a research topic and writing your dissertation proposal. This means that you need to do a great deal of reading and clear thinking to identify the problem and your area of interest. This chapter deals with choosing a topic for research and gives guidance on writing your dissertation proposal. The contents of Chapter 2 are illustrated in Figure 2.1.

Choosing the topic

When embarking upon research as part of an undergraduate or a postgraduate study programme, it is important that particular attention is paid to the choice of subject or topic of the research. The choice of topic usually comes from your interest in and value of a particular subject. This interest and value will eventually be developed into a series of questions which you are keen to find answers to. If you are finding difficulty in choosing a researchable topic, you can consider the following:

1 Consult the library catalogue and inquire about theses and dissertations, articles in academic journals, reports, books and the like.
2 Using web searching to find area of interest.
3 Talk about problems and possible topics with your colleagues and/or with your lecturers who are experts in the field.
4 Arrange an informal interview with professionals in the industry and discuss what you should emphasise and what are the possible practical outcomes.

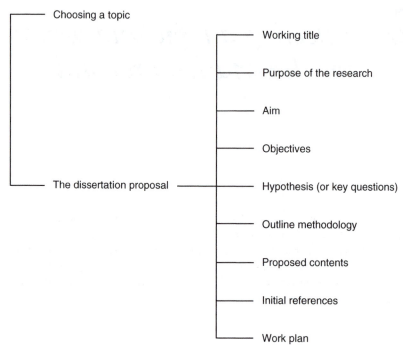

Figure 2.1   *Contents of Chapter 2*

There are a number of criteria that need to be considered when deciding on your research topic. These are summarised below:

1   *A dissertation topic should be realistic.* You need to identify a problem (supported by published materials), investigate the causes and possible solution. You might wish to investigate a problem on site, for example the causes of variation orders and its effect on the project outcome, or the problem of material management on site. Alternatively, the research subject might cover an office procedure or the appraisal of a particular system. Here you will be investigating what should be done and what is actually done.

2   *Your topic should be specific and narrow.* For example, if you are conducting a detailed case study, do not investigate all the factors that may influence site productivity as a topic, but rather the influence of financial incentives on productivity. You might be tempted to study job satisfaction in the building industry, but this again is too broad. A better way is to narrow it down and investigate job satisfaction of operatives who are working for small-size firms.

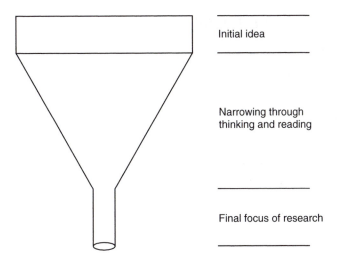

Initial idea

Narrowing through
thinking and reading

Final focus of research

Figure 2.2   *A diagram showing the narrowing down of the research topic*

Narrowing down the topic can be diagrammed as in Figure 2.2. Bouma and
Atkinson (1995) comment:

> the first thing to do if we are to narrow and clarify a problem question is to
> 'unpack' it. Most of the starting-points contain many issues and suggest many
> different avenues of research. The questions we begin with are usually quite
> complex. They may sound simple, but they are probably far from it. If we are
> to narrow and focus the issues for research, we have to list the issues involved
> in the question. We are then in a position to choose from that list, a question
> that will focus our attention on a narrowed problem.

3   *Your topic should show individuality, i.e. your personal contribution to the study.*
    What is new about your investigation? This can take the form of case stud-
    ies, a series of interviews, postal questionnaires or analysing archival data
    (Chapters 4 and 5 discuss these approaches in detail).
4   *Accessibility of information.* Your topic should be in an area that you can
    have access to the necessary data for the successful completion of the
    dissertation.
5   *Personal ability.* The subject and extent of the research should be within
    your intellectual and physical resource ability.
6   *Personal interest.* Last but not least, you should be interested in the subject
    of research. It is of no use if you do not like or enjoy the research subject.

The dissertation proposal

After deciding on a topic for research, your dissertation proposal should contain the following:

1   Working title.
2   Purpose.
3   Aim.
4   Objectives.
5   Hypothesis (or key questions).
6   Outline methodology.
7   Proposed contents.
8   Initial references.
9   Work plan.

The extent and degree of detail for your proposal should ideally take the following into account.

Working title

Working title is a short line that gives your research a 'direction' and the title might slightly change at the final stages of research. The title should give your research an identity/name which reflects the focal point and area of the research. It should therefore be clear and to the point. There are a number of questions to be borne in mind when selecting a final title. These are:

1   Does the title identify precisely what is being studied?
2   Is the title clear and concise and at the same time sufficiently descriptive to allow for rapid categorisation?
3   Has the title been stripped of superfluous words and redundancy? Phrases such as 'A contribution to ...' or 'Towards a theory of ...' are nothing more than padding.
4   Are the key nouns correctly chosen and in the proper order?

Purpose of the research (or rationale for the study)

This should be a one-page discussion of approximately 500 words which sets out (with supporting data fully referenced) the problem and the reason for the proposed study, highlighting the issues to be investigated (see Appendix 1 for examples).

Aim

Ideally a ONE sentence aim should be provided, highlighting your ultimate goal. Appendix 1 provides examples of aims but here are some specimens:

Example 1: To provide a professional guideline to engineers who are required to undertake the role of the project manager.

Example 2: To develop a conceptual model for analysing productivity barriers due to subcontracting.

Example 3: To develop a theoretical model for identifying the different factors that may influence the behaviour and effectiveness of project managers.

Example 4: To construct a comparative table showing the difference in performance between various procurement methods.

Example 5: To design a health and safety management system for small size contractors.

Note to students: Note that the above aims are phrased as if the researcher is seeking to deliver some kind of a product that would benefit the industry and other future researchers, such as providing a guideline, developing a conceptual framework, constructing a comparative table and designing a system.

Objectives

Ideally between three and five single-sentence objectives should be developed. Objectives are the breakdown of your aim (sub-aims) which focuses on finding out or establishing certain issues while achieving your aim (see Appendix 1 for examples). The objectives will then pose a number of questions which will form your research questionnaire later in the research process. As shown, try to phrase each objective in the form of:

To investigate …
To analyse …
To assess …
To examine …
To compare …
To test …
To critically appraise …
To find out …
To evaluate …
Etc. …

Hypothesis (or key questions)

If your research is designed to test/validate a hypothesis or a conceptual theory, then a one-sentence hypothesis needs to be established which should clearly and specifically state the position for the argument or investigation. A hypothesis is a tentative proposition which is a subject of verification through your investigation. Your conclusion will either support or reject your proposed hypothesis, or support part and reject others. Hypotheses can often be formulated as 'if ... then' statements or as a hunch that you have about the existence of a relationship between issues or variables. Your hypothesis should also be sharp and specific (see Appendix 1 for examples).

If your research is designed to develop a theoretical framework (as in the case of qualitative research or problem-solving approach), then a number of 'key questions' need to be formulated which should state the position for the argument or investigation (see Proposal 3 in Appendix 1).

Outline methodology

In this part of the proposal you should highlight your proposed methodology for obtaining the information necessary for the study. Research methods can take many forms (see Chapter 4). At the proposal stage, however, all you need to provide is an outline methodology, for example which academic journals you will concentrate your reading on. If you intend to interview personnel, who will they be and approximately how many of them you will interview and so on (see Appendix 1 for examples).

Initial references

Details of the core sources of references, particularly those which have been consulted in order to provide the basis of evidence and information necessary to enable the dissertation proposal to be developed, should be indicated. Each reference should provide author, title, publisher, edition and publication data (see Appendix 1 for examples).

Work plan

This should be a simple work plan or timetable – a bar chart of activities over the weeks of study is often helpful – indicating what you intend to do and when (see Chapter 10).

Summary

Selecting your dissertation topic is an important task. Therefore, you should allow a reasonable amount of time in which to develop your proposal. The first step is to identify areas which have potentials and seem interesting to you. Consult your library and read articles, books, dissertations, etc. which are related to your area of interest. Discuss your ideas with your colleagues and/or lecturers. The second step is to write your initial proposal following the instructions given in this chapter (see Appendix 1 for examples). The three most important parts of your proposal are the purpose of study, objectives and methodology. Finally, discuss your initial proposal with your appointed supervisor and make amendments, if required. Start your investigation as soon as you can and stick with your work plan.

References and additional reading

Bell, J. (1996) *Doing Your Research Project: A Guide for First-Time Researchers in Education and Social Science.* Open University Press.

Bouma, G. and Atkinson, G. (1995) *A Handbook of Social Science Research: A Comprehensive and Practical Guide for Students.* Oxford University Press.

Howard, K. and Sharp, J. (1993) *The Management of a Student Research Project.* Gower.

Naoum, S. and Coles, D. (1995) MSc Dissertation Unit Guide. Unpublished. London South Bank University.

Naoum, S. and Page, M. (1995) BSc Dissertation Unit Guide. Unpublished. London South Bank University.

3 *Reviewing the literature*

The literature review is an essential stage in conducting a research project, it amounts on average, to between 20 and 25 per cent of a dissertation content, although certain dissertations subjects may require a literature review amounting up to 50 per cent of the content. The literature review involves reading and appraising what other people have written about your subject area. It can be both descriptive and analytical. It is descriptive in that it describes the work of previous writers and it is analytical in that it critically analyses the contribution of others with the view of identifying similarities and contradictions made by previous writers.

There are five main activities involved in undertaking a literature review:

1 Knowing the sources of information.
2 Understanding how the library works.
3 Collecting existing knowledge on the subject, reading and note-taking.
4 Systematically organising the literature.
5 Appraising and writing up the literature review.

This chapter focuses on discussing the above five activities. The contents of Chapter 3 are illustrated in Figure 3.1.

Rationale for undertaking a literature review

In order to be able to make an original contribution to knowledge in your research area, the literature review should demonstrate that you have a comprehensive grasp of existing knowledge. The literature review serves two purposes. First, it seeks systematic reading of previously published and unpublished information relating to your area of investigation. The gathered information will develop issues and themes and should drive you to the next important stage, namely, research design. Second, the literature review will help you to improve your research

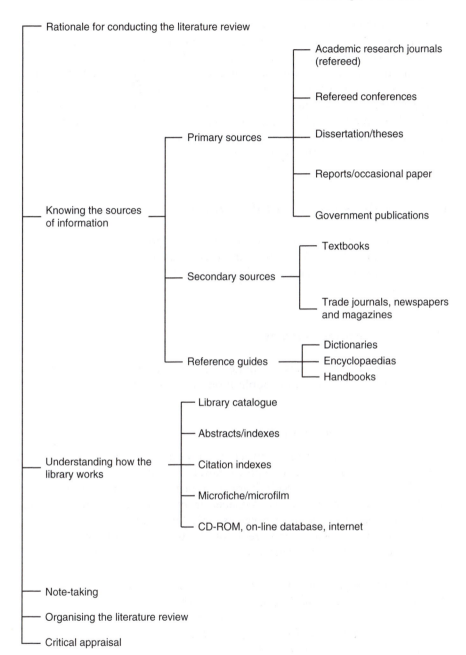

Figure 3.1   *Contents of Chapter 3*

study by looking into previous research design or questionnaires which will give you some insights into how you can design your own study more effectively. The authors of past dissertations, theses and published journal articles gave their research work a great deal of thought, and you can frequently benefit from their thinking when conducting your literature search. In other words, a literature review attempts to integrate what others have done and said, to criticise previous scholarly works, to build bridges between related topic areas, and/or identify the central issues in a field. The following sections describe the activities involved in undertaking the literature review.

Knowing the sources of information

Starting the literature review process involves the identification of appropriate literature. Generally speaking, there are three types of literature sources available for you to look at. These are primary sources, secondary sources and reference guides.

Primary literature sources

Primary literature is the most accurate source of information as it publishes original research. The following publications fall within these primary sources.

Academic research journals (refereed)

Refereed journals, related to construction, publish original research work and technical papers which usually report innovative developments in the field of Built Environment such as architecture, town planning, engineering, construction and management. By publishing papers from both industry and academia, refereed journals provide an excellent source for discussing recent developments in your field. Listed below are of some of the international refereed journals related to the built environment:

1   American Society of Civil Engineers (ASCE), New York:
    Journal of Construction Engineering and Management.
    Journal of Computing in Civil Engineering.
    Journal of Professional Issues in Engineering.
    Journal of Environmental Engineering.
    Journal of Management of Engineering.

2 *International Journal of Urban and Regional Research.*
3 *Construction Management and Economics,* Reading University.
4 *Engineering, Construction and Architectural Management,* Loughborough University.
5 *International Project Management,* Henley College.
6 *International Journal of Procurement,* University of Glamorgan.
7 *International Journal of Construction Maintenance and Repair* (now *Construction Repair*).
8 *International Journal of Construction Technology.*
9 *International Construction Law Review.*
10 *Financial Management of Property and Construction.*
11 *Economic Journal.*

See Appendix 2 for more journals.

Refereed conferences

The term 'conference' also applies to symposiums and congresses. In many fields, conference proceedings can be a main source of information and the main aim of these conferences is similar to that of academic journals (i.e. discussing current developments in your field of study). Papers published in proceedings contain a collection of themes such as mega projects, developing countries projects, productivity, design, procurement, economics, law, technology, sociology, psychology, organisation, management and many more. The majority of international conference papers are based on 'primary' research and are accepted for publication after they have been refereed by at least two members of the paper review committee. Therefore, the quality of some conference papers can be as good as refereed journal articles. The British Library Lending Division publishes an 'Index' of conference proceedings. Below is a list of some international conference proceedings related to the construction industry:

1 Organisation and Management of Construction Symposium (known as CIB W65).
2 Procurement Systems Symposium (CIB W92).
3 Building Economics and Construction Management (CIB W90).
4 Economic Management of Innovation.
5 Congress on Computing in Civil Engineering (organised by the ASCE).
6 The Design and Environment Conference.
7 International Conference on Computers in Urban Planning and Management.
8 Association of Research in Construction Management (ARCOM).

 9   Information Technology in Civil and Structural Engineering Design.
10   International Congress on Construction.
11   Automation and Robotics in Construction.
12   International Cost Engineering Council Symposium.
13   Financial Management of Property and Construction.

Dissertation/theses

It is most important for undergraduate, postgraduate and higher degree students to look at other people's dissertations or theses, where appropriate. Dissertations and theses serve two purposes. First, they enable you to have an idea about the content of the work, the standard expected, methodology adopted and the structure and style of writing up. Second, you can benefit from the list of references and bibliography that are attached at the back of the work.

Most libraries have an on-site dissertation and theses collection as well as a borrowing service through the inter-library loan. Students registered for higher degrees can obtain a list of theses from the British Library Lending Division (BLLD). Another option for finding out about other theses is to contact academic institutions requesting a list of accepted dissertations or theses. The contact is usually made through the Departmental Secretary or the Academic Thesis Supervisor. Once you have identified what you want, your library should do the rest by borrowing the material through the inter-library loan service. Requesting a dissertation or a thesis may take 4–6 weeks to arrive and overseas PhD theses may take longer.

Reports/occasional papers

Technical reports and occasional papers can be of great use to you because they are comprehensive and often publish up-to-date information. Moreover, occasional papers and reports of certain research groups are sufficiently prestigious for the researchers concerned to favour them as a method of disseminating their results (Howard and Sharp, 1993). Reports and occasional papers fall within a special category and the following institutions publish these sources of information:

1   Chartered Institute of Building (CIOB).
2   Royal Institute of Chartered Surveyors (RICS).
3   Royal Institute of British Architects (RIBA).
4   Building Research Establishment (BRE).
5   Construction Industry Research and Information Association (CIRIA).

Government publications

Government publications are one of the largest and most important sources of information, especially for those students analysing secondary data. Government documents may be classified as a) government administration records, and b) research records for specialists, including a considerable number of statistics and data of value to science and business.

Finding your way around government publications can sometimes be complicated and confusing. This is because of the sheer volume of information published under the category of government documents and because they are often placed outside the normal subject index system. For instance, most people ask for government documents by their popular name and not by their official name (The Latham Report instead of *Constructing the Team,* or the Tavistock Report instead of *Communication in the Construction Industry* and so on). Therefore, the best way to go around this problem is to ask for the monthly list or catalogue of yearly government publications of the UK, USA or elsewhere. For example, in the UK, you should be asking for the monthly lists of materials published by the HMSO or the catalogue of British Official Publications not published by the HMSO (Howard and Sharp, 1993).

Secondary literature sources

Secondary literature sources are those that cite from primary sources such as textbooks, and newspaper articles.

Textbooks

Most research is initially based on information from books which the researcher has either come across at the proposal stage or has been found in a dissertation/ thesis bibliography. Textbooks are much easier to trace and obtain than other references. However, one significant difference between books and journals is that research work published in a journal is comparatively more recent. It takes less time to have a paper published in a journal than it takes to publish a book. A second difference is that journals contain a range of articles, whereas books often deal with a particular issue. In the initial reading phase involving books, more can be gained from reading books that are edited as they are likely to contain a wide range of perspectives on a particular subject.

Trade journals, newspapers and magazines

These are mass media sources of information which often summarise research or provide views on a particular subject. It is very useful to refer to these sources at the start of your dissertation. Students often become interested in a particular topic as a result of reading an article in a newspaper, magazine or a trade journal. Below are some mass media sources that are related to the built environment:

1   *Contract Journals.*
2   *Building Magazine.*
3   *Architects Journal.*
4   *Construction News.*

However, caution should be taken in reporting the findings of mass media sources. Such sources are usually unrefereed and may easily distort or exaggerate scientific claims. Remember that most refereed journals' editorial boards practise peer review for the purpose of ensuring high-quality reporting.

Reference guides

Reference guides are very useful sources of information for short and quick answers to basic questions. The function of the reference guides is to introduce the basic information about a particular subject area. Sources that fall within these guides are dictionaries, glossaries, encyclopaedias and handbooks.

Dictionaries and glossaries

These provide a definition of terms and can be useful in helping you clarify your thoughts. Some of the preferred dictionaries are:

1   *The Great Oxford English Dictionary.*
2   *Webster's Third New International Dictionary of English Language.*
3   *The Random House Dictionary of English Language.*
4   Other specialised dictionaries are: *Kister's Dictionary Buying Guide,* the *Penguin Dictionary of Sociology* and *Dictionary of Social Behaviour and Social Research Method.*

Encyclopaedias

These give a concise description of the main aspects of a topic. The most up-to-date and authorised English encyclopaedia is the *New Encyclopaedia Britannica.* Because encyclopaedias can go out of date fairly quickly, most scientific disciplines encourage and support a variety of specialised handbooks.

Handbooks

The function of handbooks is similar to encyclopaedias, except that they are more current. Handbooks, however, have limitations in that they are written by and for other scholars with similar interests and vocabularies (Smith, 1991, p. 46). These usually assume that the user understands the methods and logic of particular disciplines, and they may not have subject indexes.

Understanding how the library works

Having identified the sources of information, the next step that you need to take is to know how to find the material. It is absolutely vital that you understand the library and how to use its sources most effectively. Bell (1996) listed three important points that you need to bear in mind while conducting a literature research:

1   Find the most relevant published materials quickly.
2   Avoid getting 'bogged down'.
3   Get into the habit of recording information derived from your reading so that it can be easily found and understood weeks, months or years later.

Therefore, you may have to spend a few days getting to know the services offered by your library, what materials it holds and where, as well as the coding or shelving system it uses (usually the Dewey decimal system). Each library should contain the following:

1   Library catalogue.
2   Abstracts/indexes.
3   Citation indexes.
4   Microfiche/microfilm.
5   CD-ROM, on-line database, Internet/World Wide Web, Telnet, JANET/NISS.

Library catalogue

This lists all the material which the library has in stock including books, journal titles, videos, etc. Around the late 1980s, most academic libraries had replaced card catalogues with computer-based catalogues. However, they both register the same details. Typical computer-screen information displays record about location of book, title of book, author's name, number of pages, year of publication and library code number. Magazines, journals and newspapers are catalogued in the same way as the books but include detailed information of which issues/volumes the library holds. Issues/volumes of journals which are not held in the library have to have a special reference to inform the user of the closest library which holds that particular journal volume.

Indexes and abstracts

Indexes and abstracts can help you to trace articles in journals. They both give the title, author and issue/volume details, but abstracts also include a short summary, often less than 200 words, of the article. An abstract gives you information about the problem and the survey findings. Indexes and abstracts not only provide a short-cut means of access to information, since they curtail endless random searching through periodical titles on the shelves, but also a large number of references which would not be found by such a haphazard method (Haywood and Wragg, 1982). Most abstracting services classify articles in accordance with the subject they cover. The natural way to start your search on a particular subject is, first, to look at the journal abstract/index. Second, study the abstract and find out whether the article is likely to be worth reading or is merely duplicating material already studied. Third, if it is worth reading, you then start searching the volume you need in order to locate the pages of the article in that volume.

Citation index

Citation indexing reports alphabetical lists by author of papers cited in published articles. This source of indexing provides standard bibliographic information on individual articles written in a particular time period. Their permuted subject index classifies each of these articles by significant words used in each article's title, and the citation index lists by author or by referenced articles.

Citation indexing is important because it graphically depicts scientific networks by quantifying who cites whom (Smith, 1991). Almost all scientific articles or reports published at any time and cited during a particular year will have an accompanying bibliographic list of authors, books, articles, reports and the like which cite the article or author. The researcher may then create a snowball sample of citations in a particular area by looking up each new list of citations. For example, if you know that Professor X is an expert on leadership, you could look up his/her name in Social Science Citation Index to start tracing a collegial network of experts.

Microfiche and microfilm

Most libraries have these facilities which means that copies of some news-papers, journals and theses are kept on film, either flat (microfiche) or more usually on reels (microfilm). This means that they take up very little room and a great deal of information can be kept in a small space. Moreover, you can read and print out the information on special readers.

CD-ROM

This is 'compact disc read-only memory'. CD-ROMs are like a music CD, but containing information. They are very important for tracing information for projects. Here, you can put in the subject you are looking for and obtain a list of relevant references. They give the information needed to trace an article, i.e. the journal you will find it in, the date it appeared, the author and title details. However, some also give the full text of the article, e.g. *The Times* and *Guardian* newspapers.

Note-taking

It is important to take notes of your literature as it is the best means to recapitu-late what you have reviewed. When taking notes, you should focus on the main issues addressed and the main arguments put forward. Dunleavy (1988) com-ments that, in conducting a literature review, one needs to get away from focus-ing too single-mindedly on what the authors are trying to say, and instead focus on what you are trying to get out of a source.

In this context, note-taking forms the basis for the construction of a literature review framework and provides the best way of maximising the extraction of information in the initial reading stage. Note-taking provides you with a focused framework and ensures that unnecessary detail is not incorporated into the review. To enable you to get the most out of the literature, it might be useful if you address each piece of literature with the following questions (Kumar, 1995):

1 What do I know already?
2 What am I looking for in this source; what do I think I need?
3 What is useful or different in this source, given my existing knowledge and interest?

It is also useful to consider a form for taking summary notes since its aim is to enable you to put these notes together in a meaningful way. It is common to find students having no difficulty in note-taking but facing considerable difficulty

Reference Details
Author(s):
Year of Publication:
Title of Article or Chapter in Edited Book:
Volume and Issue Number (Journals):
Place of Publication (Books):
Publisher (Books):
Page Numbers (Journal and Edited Book Article):
Source (Classification Number and Library Location):

1 Main Issues

2 Main Arguments

3 Research Methodology (if applicable)

4 Main Conclusions

Figure 3.2   *Sample summary note-taking form (Kumar, 1995)*

when it comes to putting these notes together to form a literature review. It can be argued that it is the way the students take the notes that causes the problem. Figure 3.2 is a sample form on which to take notes and ideally should not be more than one page in length and at the most two pages. Use one form for each piece of literature (Kumar, 1995).

Organising the literature review

Note-taking is highly important in building a good research file. The summary and abstract for each article become important for quickly sifting out the irrelevant from the relevant research. For those that appear relevant, you then proceed to read the entire text. Therefore, it is absolutely essential to start from day one of your research to maintain a literature file to store the material that you collect. If it is a four months' dissertation, one large file will probably be enough to store all the information you gather. However, whether it is a dissertation, a thesis or a research project, the principle of structuring the literature file is the same:

1 Subdivide your file into a number of topical areas. For example, suppose you are investigating the factors that affect operatives' productivity in construction sites. In this case, subdivide your manila file into a number of sections. You may create a section on definition and measurement of the term productivity, another section on previous research that investigated the factors affecting productivity rate, another on previous productivity models, etc.

2 In each of these sections you will include all the literature review that falls under the relevant topic area together with your summary note-taking form (see Figure 3.2). Such file/s will later help you to:
   (a) suggest hunches and possible research design to test these hunches.
   (b) intellectually argue the acceptance or refutation of various authors' arguments.
   (c) define key concepts and derive into logical relationships between these key concepts.
   (d) write your literature review chapters. After all, your literature chapters will contain the same topic areas as in your file/s.

3 Create an index sheet at the front of the file. This sheet will provide you with a list of all the articles, book chapters, etc. that are included in the file. Each piece of literature needs to have a separate code which will be the same code as in the index sheet. The purpose of this exercise is to provide you with a clear view as what have you covered and where.

Critical appraisal of literature review

After you have compiled the sources of information you will be ready to critically appraise the information.

A critical appraisal is analytical in nature. It critically examines the contribution of other people's work with the view to identify the following:

1 Similarities in the statements made by previous writers;
2 Common issue(s) raised by previous writers;
3 Differences or contradiction of statements made by previous writers;
4 Criticisms made by previous writers.

The following example might serve as a model to show the style of critically appraising the literature review (see pages 135–7 for another example). In this example, the critical appraisal is extracted from a dissertation by Manning (1995) entitled 'An Examination of Alternative Dispute Resolution in the Construction Industry'. You may not be familiar with the field of study, but the review puts you in the picture as what is expected from you. Manning's aim was to evaluate the success of ADR on construction related disputes to discover if it is a viable alternative to litigation and arbitration.

Example of Critical Appraisal – An Examination of Alternative Dispute Resolution in the Construction Industry.

This section of the dissertation seeks to appraise the salient points of the literature review with a view to discovering the ways Alternative Dispute Resolution (ADR) can progress in the UK, what is holding back its increased use and how the problems associated with ADR can be overcome. Finally, this section will seek to understand the role of ADR in the industry.

There appears to be a general consensus that both the Construction Industry and the system of litigation in the UK have unresolved problems [Stewart (1994), Heilbron (1993), Jones (1993)]. There is also an agreement that the main alternative to litigation in the past – arbitration – is not working either (Jones, 1993). The Civil Justice system has been described as 'Dickensian' in a Law Society report (Heilbron, 1993) and the recent Woolf Report (1995) has suggested ways of modernising the legal system by the use of new technology and faster and more efficient hearings.

The Construction Industry is at present subject to much analysis and criticism generated by the recent Latham Report (1994) which was highly critical of current procurement systems, contract terms, training systems and the whole culture of the Construction Industry in this country. The Latham Report sought radical changes to reduce conflict. The effect of this report has been the new 'Fair Construction Contracts' consultation document with a view to new legislation on construction contracts and the amendment of the New Engineering Contract, to allow its use in construction contracts (Bingham, 1995).

The move away from litigation towards a less adversarial approach should encourage an increased use of ADR in the industry. The Latham Report recommended the use of adjudication – binding for the duration of the contract. It did not actively promote the use of mediation, conciliation or mini-trials. Research in the US, where ADR is already well established as a form of dispute resolution, has shown that mediation is the most commonly used system of ADR (Henderson, 1994). To this end one might ask whether the form of adjudication recommended in the Latham Report and used in the National Exhibition Centre (NEC), is a form of ADR is debatable. ADR procedures normally involve a neutral third party and are non-binding, intending the parties to reach a settlement themselves with the aid of a neutral – adjudication seeks to impose a settlement, so it is not strictly a form of ADR.

Even though the Latham Report did not fully endorse the common forms of ADR – that is, mediation, conciliation and mini-trial, its discussion on their use and the subsequent endorsement of ADR methods in the 'Fair Construction Contracts' consultation document may help to give ADR more independent, positive publicity which could increase its future use. Obviously, ADR may not be feasible for all situations, for example, where points of law need to be decided, one party has no interest in settlement (Comes, 1994), or where one party is withholding money from another (Newman, 1993), but for some disputes it can be a viable solution.

Suitable situations for ADR include the requirement to preserve business relationships, negotiations have reached an impasse and the need for privacy (Comes, 1994). Arnold (1992) states that ADR is suitable for technical disputes where a mediator or neutral can be chosen with technical ability, whereas Newman (1993) argues that ADR is unsuitable for complex, technical disputes. It is likely that the suitability of ADR in such disputes depends on the attitude of the parties and their willingness to compromise. However, the most important prerequisite for successful ADR is the desire for the parties to explore the possibility for settlement (Comes, 1994). If this prerequisite is not present, then the ADR process will not commence at all.

The possible reason for mediation being favoured as a form of ADR over mini-trials and other forms of dispute resolution could be explained by its main advantages, speed, simplicity and low cost. The parties are also in control of proceedings, they are free to arrive at commercial decisions and explore all possible avenues of settlement (Jawad, 1994). The possibility of continued business relations being maintained in ADR is high, and in the Construction Industry this factor is particularly important (Jawad, 1994).

A possible potential for the growth of mediation in this country could be through the legal profession (Brown and Marriott, 1993). Many legal firms now contain trained mediators and it would be more in their interest to promote mediation rather than other forms of dispute resolution where their services would not be required such as expert appraisal. It is likely that an 'expert' in such cases would be a technically trained professional more along the lines of an arbitrator than a legal professional. This could be the reason mediation is the most favoured method of ADR in the construction industry.

Mediation has been criticised by some writers for becoming 'overrun' by the legal profession – Centre for Dispute Resolution (CEDR) and ADR Net are both backed by solicitors (Stewart, 1992a). A criticism of ADR in general is that it is being used as a Public Relations exercise by solicitors to promote their firms (Newman, 1992), however, this should not preclude its use, even if this is the case.

Other criticisms have focused on the firms providing the service, implying that such firms, adversely affected by the recession, are now seeking new areas for additional income (Gowan & Bolton, 1993). If ADR offers cost advantages to its users, and provides an expedient settlement, then profiting from its use may not be levelled as a criticism. Gowan and Bolton also claim that solicitors have been practising a form of ADR by negotiation – evidenced by the large number of cases settled prior to attending the court – but this is known to occur late in the process where high fees have already been incurred (Newman, 1993).

The most valid criticism of ADR is that it could provide an opportunity for 'time wasters' and those seeking to obtain information on the other side's tactics (Gowan & Bolton, 1993). Another criticism has been that there may not be adequate time to assess the details of the dispute (Naughton, 1990). While this may happen, the advantage of lower costs expended will reduce this risk. Naughton's suggestion of 'core discovery' – that is, limiting discovery to significant documents – would appear to reduce this risk (1990). An effective mediator can often detect tactics such as wasting time and seek to prevent them occurring. The success of ADR can be due to the competence of the neutral or mediator, if using the skills effectively can help to bring the sides together to reach a settlement even when they seem far apart (Naughton, 1990). Conversely, a poor mediator or neutral could have an adverse effect and can often lead to the failure of negotiations. The Construction Industry Council's suggestion (CIC, 1994) to attain a central list of trained mediators or neutrals for use in ADR could help to improve the success rate of and attitude to ADR.

Gowan & Bolton (1993) see the reform of arbitration proceedings as the answer to the problems of dispute resolution in the Construction Industry. Arnold (1992) believes the process of arbitration could be streamlined to make it cheaper and faster but as arbitration has a bad reputation at present, it could prove difficult to change this widely held belief (Stewart, 1992b). One of the problems with the use of ADR, as discussed by the CIC (1994), has been the lack of uniformity in accepted definitions – many writers define mediation and conciliation as being the same process, and others, such as the Chartered Institute of Arbitrators (CIArb), consider both processes to be different. Binding adjudication and arbitration are also considered by some writers to be forms of ADR (Bevan, 1992). The author believes, in agreement with the CIC (1994), that it would be useful to have a clear accepted definition of these terms if the knowledge about, and consequently the use of ADR is to increase.

Mackie (1991) states that, while ADR is the subject of much discussion, it is not widely used. Mackie suggests the need for increased education on alternative methods of dispute resolution and more research into the effectiveness and appropriateness of alliterative forms of dispute resolution systems. These two methods seem to be the most effective way of increasing the use of ADR. ADR should not be viewed as a panacea for all disputes, as it is clearly not suitable for all forms of dispute. Some types of dispute are suitable only for litigation or arbitration and some disputes may be resolved by negotiation between the parties themselves and do not require the assistance of a third party. ADR however can be seen as a useful adjunct to other forms of dispute resolution – some writers now refer to ADR as 'amicable' dispute resolution or 'assisted' or 'appropriate' dispute resolution (CIC, 1994).

As ADR is a confidential process, knowledge of its use is limited. It is understood that the growth in use of mediation has increased rapidly since its introduction (Jawad, 1994). The second Part of this dissertation (field work findings) seeks to determine the extent of use of ADR in the Construction Industry, its success rate, what reasons are hampering its use, what can be done to increase its use and the advantages of ADR. The results of this research are analysed and discussed in the following chapter.

Summary

The literature review is one of the earliest stages in the research process and it amounts to a significant proportion of a dissertation content. This review basically searches for material that is relevant to the subject of your dissertation with the intention to describe and analyse what has been written by others. It can take some time to get to grips with. Hence, it is advisable to allow sufficient time to cover the activities involved in undertaking a literature review. These activities are (a) knowing the sources of information, (b) understanding how the library works, (c) note-taking, (d) organising the literature review, and (e) appraising and writing up the literature review.

References and additional reading

Bell, J, (1996) *Doing Your Research Project: A Guide for First-Time Researchers in Education and Social Science.* Open University Press.

Cooper, H. (1989) *Integrated Research: A Guide for Literature Reviews.* Sage.

Dunleavy, P. (1988) *Studying for a Degree in the Humanities and Social Sciences.* Macmillan.

Haywood, P. and Wragg, E. (1982) *Evaluating the Literature.* Rediguide 2, School of Education, University of Nottingham.

Howard, K. and Sharp, J. (1993) *The Management of a Student Research Project.* Gower.

Kumar, S. (1995) Undertaking Literature Review. Unpublished paper. London South Bank University.

Manning, D. (1995) *An Examination of Alternative Dispute Resolution in the Construction Industry.* MSc dissertation. London South Bank University.

Smith, H.W. (1991) *Strategies of Social Research.* Holt, Rinehart and Winston.

Part Two

Research Design and Methodology

4   *Approaches to data collection*

Once you have determined the purpose of your study and have completed a thorough literature search, you should be ready to design your research in detail. Research design is an action plan for getting from 'here' to 'there', where 'here' may be defined as the initial set of questions to be answered, and 'there' is some set of conclusion (answers) about these questions. Between 'here' and 'there' may be found a number of major steps, including the collection and analysis of relevant data (Yin, 1994).

This chapter will review the various approaches to data collection. It first explains the difference between 'quantitative' and 'qualitative' research. Second, it explains the two main approaches to data collection known as 'fieldwork' and 'desk study', and will discuss the survey approach, the case study approach, action research and archival data collection. The contents of Chapter 4 are illustrated in Figure 4.1.

Research strategy

One of the problems of reading about research methods and techniques is the terminology. Writers use terms that may be incomprehensive to other people. In order to avoid confusion, it may be helpful to define the various terms as I go along. Here, research strategy can be defined as the way in which the research objectives can be questioned. There are two types of research strategies, namely, 'quantitative research' and 'qualitative research'. Deciding on which type of research to follow, depends on the purpose of the study and the type and availability of the information which is required.

Quantitative research

Quantitative research is 'objective' in nature. It is defined as an inquiry into a social or human problem, based on testing a hypothesis or a theory composed

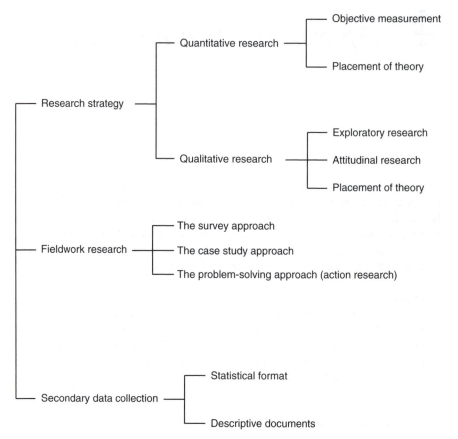

Figure 4.1   *Contents of Chapter 4*

of variables, measured with numbers, and analysed with statistical procedures, in order to determine whether the hypothesis or the theory hold true (Creswell, 1994). Quantitative data is, therefore, not abstract, they are hard and reliable; they are measurements of tangible, countable, sensate features of the world (Bouma and Atkinson, 1995).

Quantitative research is selected under the following circumstances:

1   When you want to find facts about a concept, a question or an attribute.
2   When you want to collect factual evidence and study the relationship between these facts in order to test a particular theory or hypothesis (see Proposal 2 in Appendix 1).

Chapters 6 and 7 provide examples of questionnaire construction as well as types of measurement. But here are some quantitative questions:

1   What was the programmed and the actual construction duration of the project (in weeks)?
2   What was the final cost/sq. m. of the project?
3   What method of construction did you use to build the project?
4   How many fatal accidents has your company been involved with over the past three years?

Note to students: The techniques and tools that are commonly used in data collection are described in Chapter 5.

Placement of the theory in quantitative research

In 'quantitative' study, the hypotheses, research questions and objectives can be better understood when they are grounded in a theoretical framework. The concept of a theoretical framework is explained in detail in a later section. However, at this point, I would like to discuss the term 'theory'. A theory is commonly understood to have certain characteristics. Kerlinger (1979), cited in Creswell (1994, p. 73), defined a theory as 'a set of interrelated constructs (variables or questions), that presents a systematic view of phenomena by specifying relationships among variables, with the purpose of explaining natural phenomena. Here, the systematic view might be an argument, a discussion, or a rationale that helps explain (or predict) phenomena that occur in the world.'
   A theory can be introduced as:

1   *A series of hypotheses/sub-hypotheses in the form of 'if . . . then' logic statements.* For example, a proposition that the more money you offer to operatives, the more productive they are to become, is a theory.
2   *A hunch.* For example, a hunch about the quickest method (method A or method B) to build a building of type X is a theory.

In quantitative studies one uses a theory deductively and places it towards the beginning of the plan for a study: the objective is to test or verify a theory, rather than develop it. One thus begins the study advancing a theory, collects data to test it, and reflects on whether the theory was confirmed or unconfirmed by the results in the study. The theory becomes a framework for the entire study, an organising model for the research questions or hypotheses and for the data collection procedure (Creswell, 1994). (Proposal 2 in Appendix 1 is an example of theoretically based research.)

Qualitative research

Qualitative research is 'subjective' in nature. It emphasises meanings, experiences (often verbally described), description and so on. The information gathered in qualitative research can be classified under two categories of research, namely, exploratory and attitudinal.

Exploratory research

Exploratory research is used when you have a limited amount of knowledge about your topic. Here, the interview technique is usually selected as a method of data collection (see Chapter 5). The purpose of exploratory research is intertwined with the need for a clear and precise statement of the recognised problem. Researchers conduct exploratory research for three interrelated purposes: diagnosing a situation, screening alternatives and to discover new ideas (Zikmund, 1997). For instance, personnel research managers often conduct exploratory research as a diagnostic tool to point out issues of employees' concern or to generate possible explanation for motivational patterns. For example, preliminary interviews with employees may be utilised to learn current 'hot' issues, as well as concerns about bread-and-butter issues such as wages, working conditions, career opportunities, and the like.

The raw data provided in exploratory research will be exactly what people have said (in interview or recorded conversation) or a description of what has been observed.

Questionnaire construction is discussed in Chapter 6, but the following discussion summarises the format of questions that can be asked in the exploratory study. Creswell (1994) writes: 'One typically finds research questions (not hypotheses) written in exploratory research. These research questions assume two levels (1) one or two "grand tour" questions followed by (2) no more than five to seven "sub-questions".' On the assumption that the researcher will comply with these two levels, the following ideas for a qualitative study may prove helpful:

1 Begin the research questions with the words 'What' or 'How'. Tell the reader that the study will do one of the following: discover, explain or seek to understand, explore a process, describe the experiences.
2 Pose questions that use non-directional wording. These questions describe, rather than relate variables or compare groups. Delete words that suggest or infer a 'quantitative' study, words with a directional orientation, such as affect, influence, impact, determine, cause and relate.

3 Expect the research questions to evolve and change during the study, a thought also consistent with the assumption of an emerging design. Often in qualitative studies the questions are under continual review and reformulation.
4 Use open-ended questions without reference to the literature or theory unless otherwise dictated by a qualitative design type.

The following is an example of exploratory research questions (Goodman and Adler, 1985, p. 2, cited in Creswell, 1994, p. 72):

Example – Explaining or seeking to understand

How are (these) conceptions of social studies played out – or not played out – in classroom practice? (**A grand tour question**) . . . How is each setting organised? (**The beginning of sub-question**) . . . What kind of interpersonal dynamics exist? . . . How do the students, co-operating teachers, faculty members, and pupils act? . . . What activities occur in each setting? What topics are discussed, and what information, opinions, and beliefs are exchanged among the participants?

Attitudinal research

Attitudinal research is used to 'subjectively' evaluate the 'opinion', 'view', or the 'perception' of a person, towards a particular object. The term 'object' is referred to as an 'attribute' a 'variable', a 'factor' or a 'question'. Chapters 6 and 7 provide examples of questionnaire construction as well as types of measurements. But here are some specimens of subjective questions:

1 Do you think that 'lack of worker's motivation' is a factor that can significantly affect productivity on site?
   Yes/No/Difficult to say.
2 Safety officers should have the power to stop operatives from work if they see them working in an unsafe manner, such as without helmet? Please express your level of agreement.
   1. Agree strongly 2. Agree 3. Disagree 4. Disagree strongly 5. No opinion.
3 Please indicate which of the following factors is most important to motivate operatives to work harder on site, which comes next in importance, which is third, and so forth?

Job security	1st rank, 2nd, 3rd, 4th, 5th
Money	1st rank, 2nd, 3rd, 4th, 5th
Work condition	1st rank, 2nd, 3rd, 4th, 5th
Job interest	1st rank, 2nd, 3rd, 4th, 5th
Prosperous job	1st rank, 2nd, 3rd, 4th, 5th

Note to students: The data gathered under the qualitative research can later be 'quantified' to some extent but a qualitative approach tends to value the data as 'qualitative' (Coolican, 1993).

Placement of the theory in qualitative research

In qualitative research the use of theory is less clear than in quantitative design because there is no standard terminology or rules about placement. A theory may emerge during the data collection and analysis phase of the research or be used relatively late in the research process as a basis for comparison with other theories. The placement of theory in qualitative research tends to be towards the end of the study. Therefore, the end product of qualitative research will be throwing up hunches and hypotheses which can be tested more rigorously by further quantitative research.

Creswell (1994) identifies some principles to observe about using a theory in the qualitative approach, these are: (a) employ it in a manner consistent with the type of qualitative design, (b) use it inductively so that it does not become something to test, but rather to develop and be shaped through the process of research, (c) create a visual model of the theory as it emerges, and (d) if used at the end of the study, compare and contrast it with other theories.

Comparing quantitative and qualitative research

From the above discussion you might have noticed a number of contrasting features of quantitative and qualitative research. The difference between quantitative research and qualitative research is rather like the difference between counting the shape and types of design of a sample of green houses as against living in them and feeling the environment. The difference between each one may be somehow quantifiable but such measurements will not convey the importance and the special impact of some over others.

Bryman (1988) provides a useful list of differences between the two research strategies. Table 4.1 includes some of the important dimensions. Naturally, the focal point of any research is its outcome. Although Table 4.1 shows distinctive

Table 4.1 *Some differences between quantitative and qualitative research (Bryman, 1998)*

		Quantitative	Qualitative
1	Role	Fact-finding based on evidence or records	Attitude measurement based on opinions, views and perceptions measurement
2	Relationship between researcher and subject	Distant	Close
3	Scope of findings	Nomothetic	Idiographic
4	Relationship between theory/ concepts and research	Testing/confirmation	Emergent/development
5	Nature of data	Hard and reliable	Rich and deep

features of the two strategies, the relationship between theory/concepts and research strategy in terms of verifying the theory/concept against proffering theory to emerge from the data is not as clear-cut as is sometimes implied.

Approaches to data collection

The approach to be adopted for conducting the research, depends on the nature of the investigation and the type of data and information that are required and available. For the student reading this book, there are two approaches to data collection, namely, fieldwork (primary data collection) and desk study (secondary data collection), both of which are described below.

Fieldwork research

In this book, the term 'fieldwork research' refers to the methods of primary data collection used by the researcher and must not be confused with the definition of field research as 'the study of people acting in the natural courses of their daily lives'. Here, fieldwork can be associated with three practical approaches:

1   The survey approach.
2   The case study approach.
3   The problem-solving approach (action research).

Note to students: Other approaches such as observational and experimental studies are also useful to know, but they are of very limited use to the level of students reading this book. They usually require long periods of time in the 'field' and emphasise detailed evidence.

The survey approach

Surveys are used to gather data from a relatively large number of respondents within a limited time frame. It is thus concerned with a generalised result when data is abstracted from a particular sample or population (see Sampling, page 58). There are two types of surveys available: the descriptive survey and the analytical survey.

1   The descriptive survey aims to answer such questions as: How many? Who? What is happening? Where? and When? It deals with counting the number of respondents with certain opinions/attitudes towards a specific object. The counting can be later analysed to compare or illustrate reality and trends. For example, the number of respondents in the survey that answered 'Yes' to the question, 'Should safety officers have the power to stop operatives from work if they see them working in an unsafe manner?' or the number of contractors who have more than five site accidents each year, and so on.
2   The analytical survey aims to establish relationship and association between the attributes/objects of your questionnaire, i.e. how often an attribute is associated with another attribute within the sample questionnaire. For example, you might find that most of the respondents that answered 'Yes', to the question whether 'worker's motivation affects site productivity', are operatives, and the majority who answered 'No' are site managers. In this sense, you may conclude that site managers are not aware of how important motivation can be to improve site productivity. This relationship is shown in Figure 4.2.
      In the example in Figure 4.2, the category of respondents is known as an 'independent variable' and attitude/views towards motivation is known as a 'dependent variable'. In other words, the attitude of respondents is dependent upon who the respondents are. In some surveys you might find a significant relationship between the independent and dependent variables,

Figure 4.2   *A diagram showing relationship between the independent and dependent variables*

and in others you will find no significant relationship (see Chapter 8). The terms 'dependent' and 'independent' variables are described further in the case study approach.

Note to students: There are situations when you need to explain 'why' a particular phenomenon is taking place. For example, in analysing the results of accidents on site, you may find a drop, over a period of time, in the number of accidents that the contractors have recorded in their books. You may need to explore 'why' this is happening? Is it the contractor's policies with regard to safety which made the record go down, or is it the provision of new Health and Safety regulations? The answers to these types of questions, although they can be assumed by the researcher and they are acceptable at the level of students reading this book, may not be totally adequate. Therefore, the 'why' questions and the 'cause and effect relationship' can be better explained and interpreted by designing an 'analytical case study'.

The case study approach

Case studies are used when the researcher intends to support his/her argument by an in-depth analysis of a person, a group of persons, an organisation or a particular project. As the nature of the case study focuses on one aspect of a problem, the conclusion drawn will not be generalised but, rather, related to one particular event. This is not to say that the case study approach is of limited value. On the contrary, it provides an in-depth analysis of a specific problem.

There are three types of case study designs:

1   The descriptive case study which is similar to the concept of the descriptive survey (i.e. counting), except it is applied on detailed case(s).
2   The analytical case study which is similar to the concept of the analytical survey (i.e. counting, association and relationship), except it is applied on detailed case(s).

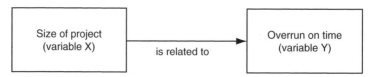

Figure 4.3   *A diagram showing the relationship between an independent variable (size of project) and the dependent variable (overrun on time)*

3   The explanatory case study which is the theoretical approach to the problem. It explains causality and tries to show linkages among the objects of the study. It asks why things happen the way they do. It also suggests that a single cause can have a specific effect. In other words, the researcher collects facts and studies the relationship of one set of facts to another, with the hope of finding some causal relationship between them.

Note to students: If the sample is large, then the relationship can be tested statistically. If the sample is small, then the relationship can only be discussed 'intellectually', which is acceptable for the level of students reading this book.

An example of an explanatory case study is when you have a hunch (related to a hypothesis or a theory) that the larger the project, the more difficult it becomes to control the construction operations and, subsequently, the project will overrun on time. This relationship/hypothesis is shown in Figure 4.3.

As mentioned earlier, analytical research means that you have identified an element that causes, affects or has an influence on another element. This is basic to the logic of a hypothesis (Bouma and Atkinson, 1995). The element that does the causing is called an 'independent variable'. The element which is acted upon, produced or caused by the independent variable is called a 'dependent variable'. In Figure 4.3, the size of project is an independent variable and overrun on time is the dependent variable.

However, time overrun is not determined solely by the size of the project. Projects can also overrun on time due to the selection of an inappropriate procurement method, client variation orders, client inexperience, environmental factors such as weather, and industrial factors such as strikes. Each of these causes will have a greater or lesser effect depending on its strength. Therefore, you need to construct a theoretical or a conceptual framework. Sometimes a model is used instead of, or interchangeably with, a theoretical framework. Using the example above, the model in Figure 4.4 is possible.

In Figure 4.4 the three variables, client experience, project complexity and size of project, are the independent variables and are related to the dependent variable, time overrun. One of the independent variables is seen to be negatively

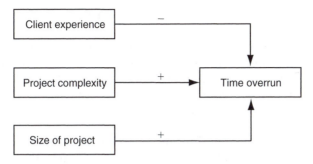

Figure 4.4 *A diagram showing three independent variables and one dependent variable*

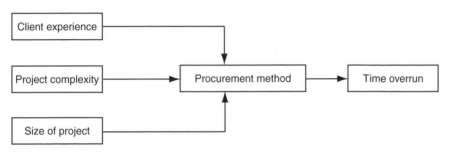

Figure 4.5 *A diagram showing three independent variables, one intervening variable and one dependent variable*

related and the other two are positively related to the dependent variable. Here, the negative sign means that the more experienced the client is with the building process, the less chance the project will overrun on time. Another possible model would be that shown in Figure 4.5.

The variables in Figure 4.5 are related in such a way that client experience, project complexity and size of project affect the selection of procurement method, which in turn affects time overrun. Here, procurement method is working as an intervening variable. An intervening variable is a process that helps to explain linkages between the dependent and independent variables and can cause the relationship between them to change.

Bouma and Atkinson (1995, p. 45) comment that the possibilities of such diagrams/models are endless. They write: 'While the most complex theories can be diagrammed, most research projects deal with only one small aspect of the whole diagram. It is often a useful discipline to diagram more than you plan to study in order to show where the proposed research fits in the larger frame of reference.'

The theoretical/conceptual framework can either:

1   Be a self-designed framework which can be formed as a result of a litera-ture search coupled with informal interviews with persons who have a wealth of knowledge about the subject area.
2   Use a previous theoretical/conceptual framework.
3   Modify an existing framework.

Note to students: There are situations, however, where the 'how' and 'why' questions can be answered by a survey and/or a case study. For example, how and why individuals within the organisation become leading managers can be studied by a survey together with a case study. The survey can examine the lead-ership pattern of organisations, showing that people with decisive characteristics are more likely to become leaders. In contrast, the case study might examine how particular leaders conduct their public relations to achieve their goal.

Problem-solving approach (action research)

With the survey and the case study approach, the researcher tends not to affect or interfere with that which is being studied. In the problem-solving approach (also named action research), the researcher reviews the current situation, identifies the problem, gets involved in introducing some changes to improve the situation and, possibly, evaluates the effect of his/her changes. This type of research is more attractive to practitioners, industrialists and stu-dents from the professional backgrounds who have identified a problem dur-ing the course of their work and wish to investigate and propose a change to improve the situation.

Examples of problem-solving research include changing organisation policy towards promotion, designing a new information flow system, recommending a new system for measuring the quality management of the organisation and the like (see Proposal 3 in Appendix 1, as an example).

It has to be stressed that whatever idea is to be recommended it must be original and practical. Moreover, the dissertation project must be structured and written as an academic piece of research and not as a project report.

Secondary data collection

The data collected using the above three approaches (surveys, case studies, problem-solving) are called 'primary' data because they are obtained first hand. While the data collected using the desk study approach are called 'secondary' data because the data are obtained from other sources. Secondary data can be stored either in a statistical or descriptive format. Secondary information has some distinctive advantages over primary data collection effort. Stewart and Kamins (1993, p. 37) note:

> The most significant of the advantages of the secondary data are related to time and cost. In general, it is much less expensive to use secondary data than it is to conduct a primary research investigation. This is true even when there are costs associated with obtaining the secondary data. When answers to questions are required quickly, the only practical alternative is to consult secondary sources. If stringent budget and time constraints are imposed on primary research, secondary research may provide higher-quality with a new research project. Secondary data also may provide a useful comparative tool. New data may be compared to existing data for purpose of examining differences or trends.

Statistical format

In construction-related research, the word 'statistical' refers to official statistics collected by the state and its agencies. These statistics are available in all public libraries and in most university libraries. The sources that publish this official information include institutions such as:

1   British Research Establishment.
2   Her Majesty's Stationery Office.
3   Chartered Institute of Building.
4   Royal Institute of British Architects.
5   Royal Institute of Chartered Surveyors.
6   Construction Industry Research and Information Association (CIRIA).
7   Building Research Establishment.

Generally speaking, the above institutions assemble the data in two ways, namely through registration or self-survey. As far as the registered information is concerned, all construction-related companies are required to provide information to the government by law. Therefore, the researcher can at any time make use of this data to show 'what is happening'. On the other hand, government bodies conduct their own routine surveys to collect data and publish them at regular intervals, say every year. For example, in the fourth proposal shown in Appendix 1, the student intended to look in depth at the type of accidents within the construction industry and compare them with other industries. He, therefore, brought together empirical evidence related to accidents on site gathered from several sources. For instance, he used the statistical data provided by the Health and Safety Executive (Statistical Services Unit) in Liverpool to analyse figures and present them in graphs. This method allowed the student to make a critical examination of the data (see p. 105 for further details on analysis).

Descriptive documents

The other method for conducting secondary data research is to analyse and critically appraise the contents of an archival document. This is similar to appraising previous literature and can include diaries, newspapers, observations, etc.

Secondary data are usually processed before the researcher starts to make use of them. Therefore, the researcher should take absolute care when using them. Below are some questions that ought to be considered when using secondary data:

1   Is the material factually accurate?
2   Is the material reliable? Would it have been the same if it had been collected by anyone else?
3   Is it systematic, providing a complete account of what it describes? What is lacking?
4   Why was it collected?
5   Is the material representative?

Of course, there is no way that the researcher can answer any of these questions with certainty, but they should be borne in mind. Nevertheless, most of these questions have to be asked when collecting primary data as well.

After gathering the information that is required for your investigation you will then need to analyse it. Chapter 8 explains various methods of data analysis.

Summary

This chapter reviewed the various approaches to data collection in construction. It first distinguished between quantitative and qualitative research. Quantitative research is an objective measurement of the problem. It investigates facts and tries to establish relationships between these facts. While a qualitative research is a subjective assessment of a situation or problem, and takes the form of an opinion, view, perception or attitude towards objects. An object is referred to as an attribute, variable, factor or question.

This chapter also explained the approaches to data collection that are suitable for the level of students reading this book. It explained the survey approach, the case study approach, action research and archival data collection approach.

References

Bouma, G. and Atkinson, G. (1995) *A Handbook of Social Science Research: A Comprehensive and Practical Guide for Students.* Oxford University Press.

Bryman, A. (1988) *Quantity and Quality in Social Research.* Unwin Hyman.

Coolican, H. (1993) *Research Methods and Statistics in Psychology.* Hodder & Stoughton.

Creswell, J. (1994) *Research Design: Qualitative and Quantitative Approach.* Sage.

Stewart, D. and Kamins, M. (1993) *Secondary Research: Information sources and methods,* 2nd edn. Sage.

Yin, R. (1994) *Case Study Research: Design and Methods.* Sage.

Zikmund, W. (1997) *Business Research Methods.* Dryden Press.

5   *Techniques for data collection*

After you have decided on the type of data to be collected (quantitative, qualitative or both), and have also decided on the research approach (survey, case study, action research or analysing archival data), you will be in a position to think seriously about the technique for collecting the data. This chapter will explain the main features of two major research techniques available to elicit data and information from respondents. These are the postal questionnaire and the personal interview. It has to be stressed, however, that data collection does not necessarily depend on one method or technique. Some research depends solely on one method of data collection but others do not. In other words, it is possible for you to conduct a postal questionnaire and a case study, or to conduct interviews in addition to a postal questionnaire, or whatever. It is also possible that your dissertation questionnaire includes quantitative questions as well as qualitative ones. Your decision will depend on your judgement as to which method or techniques will best obtain the information you need in order to achieve the purpose of your study. Whatever decision you make you need to consult your supervisor before going ahead with it. The contents of Chapter 5 are illustrated in Figure 5.1.

Postal questionnaire

The postal questionnaire is probably the most widely used data collection technique for conducting surveys. It is most suited to surveys whose purpose is clear enough to be explained in a few paragraphs of print, in which the scheme of questions is not over-elaborated. Postal questionnaires have been widely used for descriptive and analytical surveys in order to find out facts, opinions and views on what is happening, who, where, how many or how much. Almost all postal questionnaires have 'closed-ended' questions that require a specific response such as 'yes' or 'no' or ranking the importance of factors (see Chapter 6 for details).

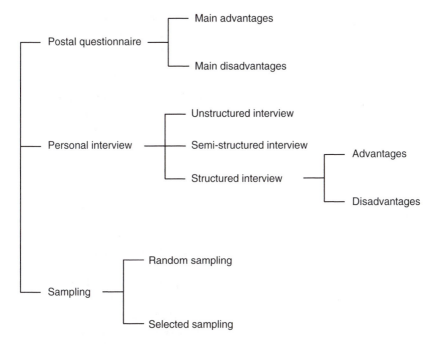

Figure 5.1 *Contents of Chapter 5*

The main advantages of postal questionnaires are:

1   *Economy.* Postal questionnaires are perceived as offering relatively high validity of results because of their wide geographic coverage. As a result, it is more suited to assembling a mass of information at a minimum expense in terms of finance, human and other resources.

2   *Speed.* Postal questionnaires are certainly a quick method of conducting a survey. If administered properly, the bulk of the returns will probably be received within two weeks. However, time must be allowed for late returns and responses to follow-up attempts. For example, if there is a lack of response to the first return of replies (beginning about two weeks after initial posting), a reminder needs to be sent to those who have not yet returned the questionnaire. Therefore, a period of about four weeks needs to be allowed in your programme of work from the date of the initial mailing to the commencement of the final analysis.

3   *Consultation.* In certain cases respondents may not have the information to hand, particularly when the questions are of a quantitative nature, and

may need to consult a document or a colleague in order to give accurate answers. Examples are questions like: 'When was the project built and what was its building cost?' or 'How many fatal accidents has your organisation been involved in over the past five years?'. Such questions may have to be answered in the respondent's own time, rather than provided on the spot as is usually associated with interviews. However, such a problem can be overcome in the interview if the questionnaire is posted to the respondent in advance of the interview.

The main limitations of postal questionnaires are:

1 *Must contain simple questions.* The postal questionnaire is only suitable for simple and straightforward questions which can be answered with the aid of easy instructions and definitions. The questions should be very carefully worded and free from faults such as ambiguity, vagueness, technical expressions, difficult questions and so forth. These faults can affect the results of the postal questionnaire even more seriously than when conducting an interview (see the following section, Personal interview).
2 *Inflexible technique.* Inflexible in the sense that postal mail questionnaires do not allow the opportunity for probing. In other words, the answers have to be accepted as final and there is no opportunity to clarify ambiguity or to appraise the non-verbal behaviour of respondents, though the latter can sometimes create bias.
3 *Accuracy.* Respondents may answer generally when you are seeking a response on a specific level of analysis. People may also answer according to what they think you want to hear. They may answer according to their public profile rather than the underlying corporate reality.
4 *No control over respondents.* This means that although you state in your questionnaire that a particular person should complete the questionnaire (such as marketing director, managing director or the site agent), there is no guarantee that this statement will ensure that the right person completes the questionnaire. However, this is less of a problem than not getting a response at all, given the fact that response rates for postal surveys usually range between 40 and 60 per cent.
5 *Industry fatigue.* Companies receive a steady stream of questionnaires and the pressures of modern business mean that for many organisations and individuals, students' questionnaires are of less priority.

Note to students: Some of the limitations of the postal questionnaire can be overcome by supplementing it with personal interviews.

Personal interview

The personal interview is another major technique for collecting factual information as well as opinions. It is a face-to-face interpersonal role situation in which an interviewer asks respondents questions designed to elicit answers pertinent to the research hypothesis. The questions, their wording and their sequence define the structure of the interview (Nachmias and Nachmias, 1996). The interview technique is suitable under the following circumstances:

1 When the people being interviewed are homogenous and share the same characteristics.
2 When you know enough about your interviewee so you only ask what is important and know how to ask it.
3 When interpersonal contact is essential to explain and describe the questions.
4 When a case study needs to be investigated in detail asking questions such as how and why things had happened the way they did.
5 When the research requires an explanation as why the respondents are answering or feeling the way they do, i.e. requires more than a Yes or No, or Agree or Disagree answer.

Interviews can take three forms, unstructured, structured and semi-structured. Some research may require one form of interview, while others may require a combination of the three forms. There are many other terms in use to distinguish between what are called here structured and unstructured interviews. For example, there are formal and informal, inflexible and flexible, standardised and unstandardised, controlled and uncontrolled interviews.

Unstructured interview

This form of interview uses 'open-ended' or 'open' questions and the questionnaire is often pitched at a very general level so that the researcher can see in what direction the interviewee takes things in their response. It is usually conducted with qualitative research (see Qualitative research in Chapter 4).

Unstructured interviews can also be conducted at the beginning of any research (also known as exploratory interviews) when the researcher knows little about his/her subject area. However, in this type of interview you need to have a clear research outline that you are familiar enough with to carry the general points around in your head. Here, there is no set order or wording of questions, no schedule and you are not looking for the same information from each person. It is purely an exploratory exercise. For example, suppose you are studying the factors that affect job satisfaction for construction operatives and you wish to explore your thinking around this research outline through informal interviews. On your first interview, the operative may focus his/her discussion around salary, job security and promotion as motivating factors affecting his/her satisfaction with the job. On your informal interview with the second, third or fourth operatives, you will seek to find out whether the same factors apply to them as well, and if there are other factors that need to be added. For instance, other operatives may add leadership, working environment and the organisation itself. At the end of the unstructured interviews you will probably end up with a list of ten factors which you will then examine on a wider sample through structured interviews or a postal questionnaire.

Note to students: Some students rely solely on two or three exploratory unstructured interviews which is not enough from which to make firm conclusions.

Semi-structured interview

This is more formal than the unstructured interview in that there are a number of specific topics around which to build the interview. This form of interview uses 'open' and 'closed-ended' questioning but the questions are not asked in a specific order and no schedule is used. Your task is to discover as much as possible about the specific issues related to your subject area. Merton and Kendal (1946) state four distinguished characteristics of the semi-structured interview or, as they named it, a focused interview:

1   It takes place with respondents known to have been involved in a particular experience.
2   It refers to situations that have been analysed prior to the interview.
3   It proceeds on the basis of an interview guide specifying topics related to the research hypotheses.
4   It is focused on the respondents' experiences regarding the situations under study.

Semi-structured interviews start by asking indirect questions in order to build up a rapport with the respondent and then explore the specific issues that the interviewer has in mind. Suppose you are investigating the factors which cause conflict between managers and the site agents. You may start by asking the site agents 'What sort of problems do site agents have in getting along with their managers?' and 'What sort of disagreement do you have with your site manager?'. You can then move on to ask more formalised questions such as

'Have you ever had any disagreement with your manager over the following: (a) allocation of resources, (b) scheduling, (c) overtime pay, (d) promotion?'

In the semi-structured interview, the interviewer has a great deal of freedom to probe various areas and to raise specific queries during the course of the interview. For example, the first question above can be probed by asking 'In what way do site managers try to restrict your work?' or 'Do any of your colleagues have the same problem?' and the like.

Structured interview

In the structured interview, questions are presented in the same order and with the same wording to all interviewees. The interviewer will have full control on the questionnaire throughout the entire process of the interview. In this technique the questioning may start with some 'open' questions, but will soon move towards a 'closed' question format. Nachmias and Nachmias (1996) cited three assumptions of the structured interview:

1   That for any research objective, the respondents have a sufficiently common vocabulary so that it is possible to formulate questions which have the same meaning for each of them.
2   That it is possible to phrase all questions in a form that is equally meaningful to each respondent.
3   That if the meaning of each question is to be identical for each respondent, its context must be identical and, since all preceding questions constitute part of the context, the sequence of questions must be identical.

The main advantages of the structured interview therefore are:

1   The answers can be more accurate.
2   The response rate is relatively high (approximately 60–70 per cent), specially if interviewees are contacted directly.
3   The answers can be explored with finding out 'Why' the particular answers are given.

Note to students: There is another data collection technique which is known as the 'telephone interview'. The rules and techniques that are applied to telephone interview are similar to face-to-face interviews. It incurs high cost but it is a quicker way to find a respondent to interview. It may not be as detailed as face-to-face interviewing, however, technological changes and improvements in communication equipment have made telephone interviewing easier, especially when researchers use random-digit dialling and computer-assisted telephone interviewing (see Nachmias and Nachmias, 1996).

Sampling

Once you have decided on the technique for collecting your fieldwork data and have thought about what you want to ask, you should be ready to decide on the characteristics of the respondents. In all cases a sample has to be drawn from its population. The term 'sample' means a specimen or part of a whole (population) which is drawn to show what the rest is like. For instance, if you are conducting a postal survey then you need to obtain a list of respondents which is known to fall into the category of your survey (survey sample). If you are conducting case studies interviews then a relative small number of cases (interview sample) is used for inferences to all cases (a population).

Selecting the research sample is very important and great care must be taken when choosing the type of sample design. The researcher has to ensure that the characteristics of the sample are the same as its population and act as representative of the population as a whole. Usually, the means of drawing a representative sample is done either randomly or non-randomly. The term 'random' means selecting subjects (respondents) arbitrarily and without purpose.

Designing the research sample can take many forms, each of which is suitable to a particular situation. Listed below are some of the types of sampling designs available to the researcher:

1   Non-random accidental sampling.
2   Non-random purposive sampling.
3   Simple random sampling.
4   Systematic random sampling.
5   Stratified random sampling.

Detailed description of the above types of sampling design is beyond the scope of this book. Students conducting an MPhil or PhD research should know about these types and ought to consider reading the appropriate parts of the reading references on sampling procedures provided on page 62. However, for the purpose of this book, you need to know that there are two types of sampling: random sampling and selected sampling.

Random sampling

This type of sampling can be used when specifics about the characteristics of the sample are not essential, such as background of respondents, size of company and type of work, etc. However, there are two main criteria that you need to take into consideration when selecting your sample. First, what do you want to know? Second, about whom do you want to know it? Both of these questions can be answered by referring back to the purpose of your study, i.e. your aim, objectives and hypothesis (or key questions).

In order to draw a random sample you need to follow two steps:

1 Identify the population from which the sample is to be drawn. This means obtaining a list of names and addresses of say, top contractors operating in the Greater London area. Such names and addresses can be obtained from your direct personal contacts or from organisations and references such as the Building Federation, Chartered Institute of Building and the Royal Institute of British Architects.
2 If the list is small, you may be able to send your questionnaire to all the names identified in your list. If the list is long, you then need to devise a method of random selection which ensures that each subject (in this case contractor) has the same probability (chance) of selection. In other words, adopt a random numbering technique to select a sample. For example, if you have a list of 400 contractors and you want a sample of 80, you might select every fifth name on the list.

Selected sampling

This type of sampling is usually chosen with the interview approach. It begins by choosing a list of names and addresses of participants with specific characteristics, for example the top contractors who are offering alternative procurement methods and undertaking refurbishment work for commercial clients. All other contractors will be excluded from the survey. Consider the study by Naoum and Langford (1987) entitled 'Management contracting: the client views'. A structured interview questionnaire was used to obtain the clients' view of management contracting and to compare the management contracting system with the traditional form of contract, having in mind clients' needs in terms of function, cost, speed and aesthetics. This section is not intended to be a thorough treatment of questionnaire construction and administration, as this is covered in Chapter 6. However, it is important to show the questionnaire that Naoum used when interviewing the clients in order to follow the discussion. Table 5.1 provides some of the questions asked. The following procedure was used by Naoum when selecting the ten clients:

1 The name of clients and their addresses were identified from the management contractors who were interviewed in phase one of the research.

Table 5.1   *Question to clients on management contracting (MC) versus traditional contracting*

Questions	Yes	Same	No
Is MC more risky to clients? Is MC more profitable to the contractor? Does MC involve fewer claims? Is MC more flexible to variations? Does MC allow an earlier start on site? Is MC quicker during construction? Is MC more reliable in predicting the build time? Is MC cheaper? Is MC more reliable in estimating the building cost? Does MC provide more control over subcontractors? Does MC exercise more control over construction tasks? Does MC provide a better building design?			

2   Names of persons who were suitable to participate in the research were identified through telephone conversations with the client organisations.
3   The ten clients selected for the interviews were homogenous and shared similar characteristics. They were all large organisations, were very experienced with the building process and, most importantly, they all undertook different types of work under different contractual arrangements. There were, however, differences in that some clients were private organisations and others were public agencies. Moreover, due to the nature of their business, some clients had complex organisational structures while others had simple ones.
4   A structured interview questionnaire was designed and posted prior to the interview.
5   The interviews provided some freedom to probe certain areas and raised specific queries during the course of the interview. More specifically, 'why do you feel that way?' and 'why do you hold that view?' or 'how did this happen?'.

In this study, a selected sample of clients was thought to be more appropriate than a random sample due to the nature of the study. If the interviews are based on people who are very different in characteristics, philosophy and experience, the questions will have different meanings for each of them. The answers, therefore, may be ambiguous for the subject under investigation. However,

Table 5.2 *Comparison between a postal survey and interview technique*

Features	Interviews	Postal questionnaire
1 Identity of respondents.	Known.	Unknown.
2 Interaction between interviewer and respondent.	Close.	Distance.
3 Time involving the researcher.	Long time to go through the interview.	Short time.
4 Cost.	High.	Significantly lower than the interviews.
5 Sample.	Small.	Large.
6 Quality of information.	Deep and detailed.	Rich.
7 Skill and experience.	The interviewer needs to have the skill to ask questions and, if necessary, to probe.	No skill required.
8 Control of the process.	High.	Low.
9 Flexibility.	Allows great flexibility to reword questions and clarify terms that are not clear.	Rigid. The answers are accepted as they are.
10 Analysis of the results.	Difficult and become complicated in the unstructured interviews.	Easy to analyse.
11 Interviewer bias.	The flexibility of interviews allows for bias. Sometimes the non-verbal communication or behaviour of the interviewee may mislead the interviewer to incorrect judgement.	If sample is selected appropriately, there should be no bias.

this does not mean that you are not permitted to have a sample of different characteristics. Your research design may require you to compare the views of two groups of people who have different backgrounds. For example, comparing the views of quantity surveyors and clients on the concept of value management technique.

Summary

This chapter discussed the main features of the postal survey and the personal interview together with showing the advantages and disadvantages of both techniques (see Table 5.2 for summary). The postal questionnaire is an impersonal survey technique that is most suitable for a descriptive or analytical survey. Its main advantages are cost-effectiveness, speed of process and that respondents can consult others for information. Its main limitations are that it must contain simple questions, it is inflexible and has no control over respondents.

The personal interview is a face-to-face situation that is most suitable for case study research and studies that require respondents with homogenous characteristics. Interviews can take three forms: unstructured, semi-structured and structured. The main advantages of personal interview are knowing the identity of respondents, interaction between interviewer and respondent, that it allows probing, the quality of information is deep and detailed and the interviewer has a high level of control over the interview process. The main limitations of the personal interview are high cost, long process time, the sample can be small and it can be difficult to analysis the information.

References and additional reading

Barnett, V. (1991) *Sample Survey Principles and Methods*. Arnold.
Bouma, G. and Atkinson, G. (1995) *A Handbook of Social Science Research: A Comprehensive and Practical Guide for Students*. Oxford University Press.
Jolliffe, F. (1986) *Survey Design and Analysis*. Ellis Horwood.
Merton, R. and Kendal, P. (1946) The focused interview. *American Journal of Sociology*, **51**, 541–557.
Moser, C. and Kalton, G. (1985) *Survey Methods in Social Investigation*. Gower.
Nachmias, C. and Nachmias, D. (1996) *Research Methods in the Social Sciences*, 5th edn. Arnold.
Naoum, S. and Langford, D. (1987) Management contracting: the client views. *American Society of Civil Engineers. Journal of Construction Division*, September.
Stewart, D. and Michael, K. (1993) *Secondary Research: Information sources and methods*, 2nd edn. Sage.

6 Questionnaire construction

Chapter 5 introduced several techniques for gathering information. As discussed, the selection of the technique depends largely on your choice of the research approach. For instance, if your research requires you to conduct case studies, then the interview technique is more appropriate than the postal questionnaire, and if your study seeks to survey the opinion of top contractors operating in London, then the postal questionnaire will be more appropriate and feasible than interviews. After deciding on the technique, your next step is to construct your questionnaire.

The foundation of all questionnaires, whether to be mailed or to be used for interviewing, is the questions. At first, the order and wording of the questions are not crucial. Your aim is to list the ideas which can be developed into a questionnaire at a later stage. The principle that you need to follow in constructing your questionnaire is shown in Figure 6.1.

Figure 6.1 shows that, during the research process, the research objectives are translated into specific questions. Therefore, while thinking what item or question to include in your questionnaire, you should ask yourself the following:

1   Which objective is the question related to?
2   Is the question relevant to the aim of the study?
3   Is the question relevant to the research hypothesis?
4   Can the answer be obtained from other sources?

This chapter describes the stages for questionnaire construction and administration. It describes and illustrates (through examples) the types of questions, question format and typical questionnaires. It also provides criteria for writing the questions. The contents of Chapter 6 are illustrated in Figure 6.2.

Note to students: A typical questionnaire format can be seen in Appendix 3.

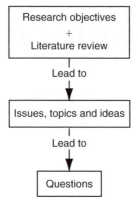

Figure 6.1   *A diagram showing that during the research process, research objectives are translated into specific questions*

How to construct a questionnaire

There are three fundamental stages that you should take in constructing your questionnaire:

1   Identifying the first thought questions.
2   Formulating the final questionnaire.
3   Wording of questions.

Stage 1: identifying the first thought questions

Before constructing you questionnaire, you should go back to your proposal and the literature file, and start formulating the 'first thought' list of questions. At this stage, the order and wording of the questions are not crucial. Your aim is to write down all possible questions (say, 20–30 questions) which are related to your research (you will edit and order them later). Suppose you are investigating whether the characteristics of site managers have an effect on project performance. You will be thinking of asking certain types of questions and not others. Whatever questions you intend to ask, they should not be arbitrary and need to be based on your literature review. For example, you may not include in the first thought questions a question about the site manager's favourite

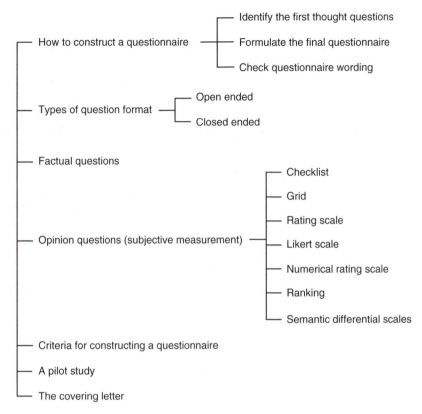

Figure 6.2 *Contents of Chapter 6*

location of work, but you are expected to include a question about his/her educational achievement. The reason for this is that you are either assuming that location of work has nothing to do with time overrun, while educational achievement has, or because previous studies in other fields have supported the second assumption. Moreover, should you ask a question about the site manager's personal life? This may just be relevant. But if you start to cover every aspect, then you will end up with a very lengthy questionnaire which could be rather hopeless. Therefore, it is absolutely important that, at the proposal stage, you need to have the purpose of your study specific, narrowly defined and stated clearly. This should provide you with a more focused research and help you to formulate the final questionnaire more easily.

Stage 2: formulating the final questionnaire

After you have identified the first thought questions, you will be ready to construct your final questionnaire. To do that, you need to introduce a number of sections or categories for the questionnaire (say, four to six sections) and try to fit the first thought questions in these sections. Give these sections a title or a theme which should correspond closely with the objectives. For example, in Proposal 1 (in Appendix 1), there were two objectives. The first objective was to survey large private house building companies and investigate the company business philosophy, organisational structure, marketing information and their marketing policy. The second objective was to relate the companies' marketing philosophy with their financial performance. From these objectives, the student began to think of what to ask and what not to ask. He constructed the questionnaire with six sections to correspond with his objectives. The first section of the questionnaire was named 'General Information'. The second section was named 'Overall Business Philosophy' and, here, he listed six issues that he thought important and relevant to ask. These six issues were:

1   Type of business.
2   Business philosophy.
3   Management philosophy on products.
4   Marketing plan.
5   Management philosophy on the market.
6   Management success criteria.

These items or ideas were not introduced arbitrarily but were based on his background literature review (see Table 6.1).

Stage 3: wording of questions

After deciding what to ask, you should then check how you have asked them. The last section of this chapter gives details on criteria for constructing the questionnaire, but here is a summary:

1   Your questions/questionnaire should be short but comprehensive.
2   Avoid leading questions.
3   Avoid double questions.
4   Avoid presuming questions.
5   Avoid hypothetical questions.

Table 6.1   *Questionnaire related to marketing*

Please respond to the following questions by ticking beside the option (letter) which best fits your company

SECTION – OVERALL BUSINESS PHILOSOPHY

1   **What business are you in?**
   (a)   We build houses.
   (b)   We build homes.
   (c)   We market homes.

2   **In the house-building cycle, what action best describes your firm's business philosophy?**
   (a)   We first determine customers' wants; and then design the product types and packages to satisfy those wants.
   (b)   We first start to build the estate; and then work out how best to sell the product.

3   **What is the management's philosophy on products and markets?**
   (a)   Management primarily thinks in terms of selling current and new products to whoever will buy them.
   (b)   Management thinks in terms of serving a wide range of markets and needs with equal effectiveness.
   (c)   Management thinks in terms of serving the needs and wants of well-defined markets chosen for their long-term growth and profit potential for the company.

4   **Does management develop different offerings and marketing plans for different segments of the market?**
   (a)   Yes.
   (b)   Sometimes.
   (c)   No.

5   **Which of the following best describes top management's philosophy guiding the firm's current marketing effort?**
   (a)   Customers will favour those products that are affordable and therefore the main task of management is to pursue improved production?
   (b)   Customers will not buy enough of the organisation's products unless the organisation makes a substantial effort to stimulate their interest in its products.
   (c)   The key of achieving organisational goals consists of the organisation's determining the needs and wants of target markets and adapting itself to delivering the desired satisfactions more effectively than its competitors.

6   **Top management's focus is on ____ (please print a, b or c)?**
   (a)   Cost (i.e. how to do things cheaply and well).
   (b)   Sales volume (i.e. maximise sales targets).
   (c)   Profit (i.e. maximise profit).

6   The questions must not be ambiguous.
7   The questions should be logical in their sequence.
8   The questionnaire must be attractive in appearance.

Note to students: Examining other questionnaires to identify possible clues through which to construct yours, is essential at this stage. This means that you should look at good questions or questionnaires in books, journals, dissertations, theses, etc. and try to copy the ideas. Remember, previous authors of past work have given their research a great deal of thought, and you can frequently benefit from their thinking when designing your questionnaire. For instance, question 3.2 of section (three) in Appendix 3, is a modification of a similar research which was conducted in the manufacturing industry. The student fine-tuned the wordings to suit his field of investigation. Provided that you make a reference to other peoples work, you will not be penalised.

Types of questions

Questionnaires are classified into two types: the 'open' form or unrestricted type, and the 'closed' form or restricted type. Open questions seek to encourage the respondent to provide free responses. For instance, the question 'What do you personally feel are the most important factors that make workers work harder?' is an open-ended question designed to study people's opinion. Oppenheim (1996) notes:

Open or free-answer types of questions are not followed by any kind of choice, and the answers have to be recorded in full. In the case of a written questionnaire, the amount of space or the number of lines provided for the answer will help to determine the length and fullness of the responses we obtain. Inevitably, some of this richness is lost when the answers are classified later, but it is useful to report a few such answers in the final report to give the reader some of the flavour of the replies. Statistical tabulations are important and must remain our first aim, but they make dull reading.

The main advantages of the open type of questions are therefore:

1   They give the respondent the opportunity to express their views. Once the respondent understands the theme of the investigation, he/she can let the thoughts roam freely, unencumbered by a prepared set of replies.

2 They are easy to ask.
3 They are more appropriate to construct with interview questionnaires, but can also be used in postal surveys. During the interview, the interviewer has a great deal of freedom to probe various areas and to raise specific queries during the course of the interview.
4 They are useful to adopt when sensitive information is required from the respondent.

Open questions can, however, present problems. As the open question offers no direct clues and is broad based, predictably this type of questionnaire is more difficult to analyse and interpret. Therefore, some form of content analysis is required for open question (see Chapter 8).

By contrast, closed questions often require a short response in the form of Yes or No, Agree or Disagree, Important or Not Important, etc. Closed-ended questions are easy to ask and quick to answer, they require no writing by either respondent or interviewer, and their analysis is straightforward (Nachmias and Nachmias, 1996). Their major drawback, however, is that they may introduce bias, either by forcing the respondent to choose from given alternatives or by offering the respondent alternatives that might not have otherwise come to mind.

The appropriateness of either a closed-ended or open-ended question depends on a number of factors. Paul Lazarsfeld (1944) suggested the following considerations to determine appropriateness of which type of question to ask:

1 The objective of the questionnaire.
2 The respondent's level of information about the topic in question.
3 The extent to which the topic has been thought through by the respondent.
4 The ease with which respondents can communicate the content of the answer or the extent to which respondents are motivated to communicate on the topic.

Factual questions

This chapter has so far discussed the type of questions. The rest of the chapter will explain and illustrate the different formats for asking the questions. I will start with the factual questions.

Factual questions are mostly designed to elicit information related to the following:

1 Facts related to the background of an individual or organisation. Most questionnaires, whether for descriptive or analytical purposes, almost certainly begin by asking factual questions such as:
   (a) What is your position within the organisation?
   (b) How many professional employees work within the organisation?
   (c) Which type of projects is your firm specialised at?
   (d) What is the annual turnover of your organisation?
2 Facts related to events or projects. For instance, the questionnaire in Table 6.2 contains questions which are designed to find facts about the characteristics of projects and their performances. In this example, the researcher's aim was to investigate the difference in performance of projects delivered by a management contract and the traditional form of contracting. A theoretical model was formulated and a number of sub-hypotheses were proposed. In order to test the sub-hypotheses, factual data was collected from previously completed projects and the performance of these projects was measured objectively.

Note to students: It is extremely difficult to determine causality in such studies unless a detailed case study is undertaken or a controlling variable system is used. This means you ought to be selective in your sample. For example, you may need to study certain categories of clients or projects and exclude others. In this way you will be limiting the variables that determine causality. Nevertheless, at the level of students reading this book, if some kind of relationship is established between the variables, then that should be sufficient. For MPhil/PhD level an in-depth investigation to causality is necessary, for example, follow-up interviews should be conducted with those involved in the study to explain how and why certain events took place. Otherwise, causality can be best studied in an experimental type of research such as laboratory and social behaviour research. As mentioned earlier these are outside the scope of this book. They basically involve validating the hypotheses under real test.

3 Facts gathered from secondary data. As mentioned in Chapter 5, secondary data are available in two formats: statistical and descriptive. For example, in the Proposal 4 shown in Appendix 1, the student examined and analysed the type of accidents within the construction industry using material which was published by the HSE Statistical Services Unit in

Table 6.2 *Project details questionnaire*

SECTION 1 – QUESTIONS RELATED TO PROJECT CHARACTERISTICS

1 What was the project type? (Please tick the appropriate option)
- (a) Housing
- (b) Industrial
- (c) Commercial
- (d) Other type (please specify)

2 What was the type of construction?
- (a) Steel structure frame
- (b) Reinforced concrete
- (c) Both (a) and (b)
- (d) Other (please specify)

3 How was the project funded?
- (a) By private funds
- (b) By public funds
- (c) By both funds
- (d) By other funds (please specify)

4 What was the gross floor area in square metres? (Please state)

5 Was it a new build project, refurbishment or both? (Please state)

6 How many main subcontractors were involved in the project? (Please state)

SECTION 2 – QUESTIONS RELATED TO THE CONTRACTUAL ARRANGEMENTS

7 What procurement method was selected for the project?
- (a) Traditional
- (b) Management contracting
- (c) Design and build
- (d) Project management
- (e) Other method (please specify)

8 What was the tendering arrangement?
- (a) Negotiation
- (b) Open tendering
- (c) Selective tendering

Table 6.2   *(continued)*

    (d)   Two-stage selective tendering
    (e)   Serial or continuous
    (f)   Other method (please specify)

9   **What was the tendering document?**
    (a)   BOQs/drawings
    (b)   Drawings/specification
    (c)   Approximate BOQs
    (d)   Other document (please specify)

SECTION 3 – QUESTIONS RELATED TO PROJECT PERFORMANCE

10   **Please give the overall contractual price tender accepted for the building in £s.**

11   **Please give the actual price on completion of building in £s.**

12   **If the actual cost exceeded the estimated, was it authorised by the client?**
    (a)   Yes
    (b)   No

13   **If your answer to question 12 above is Yes, how much of the amount was authorised by the client?**
    (a)   All of the amount
    (b)   Some of the amount (please state amount)

14   **Please give the programmed duration of the project (in weeks).**

15   **Please give the actual duration of the project (in weeks).**

16   **If the actual duration exceeded the programmed, was the extension authorised by the client?**
    (a)   Yes
    (b)   No

17   **If your answer to question 16 above is Yes, how many weeks were authorised by the client?**
    (a)   All extension requested
    (b)   Some of the requested extension (please state)

Liverpool (see the HSE record in Appendix 5 as an example). From these statistics, the student compared the accident rate of the construction industry with four other major industries within the UK. He also took the accident statistics for the construction industry and separated them into their elements and components to show their type and relation to employees. Analysis of these factual records was then presented in graphical forms (see pages 105–106).

Opinion questions (subjective measurement)

There are several formats in which opinion type of questions can be asked:

1 Checklist.
2 Grid.
3 Rating scale.
4 Likert scale.
5 Numerical rating scale.
6 Ranking.
7 Semantic differential scales.

Checklist

This type of question is essentially a list of items, about themselves, an organisation or an event, that respondents are offered to mark or tick. It is a straightforward means of collecting information and the data can be analysed easily. For example, a question may be a list of management techniques and the respondent is asked to tick which of these techniques are used by his/her organisation. It may be a list of motivational factors and respondents are asked to tick which of these apply to them, and so on.

The checklist questions are specially designed for a group of respondents who have accurate information and can answer the questions with a high degree of certainty. For example, a respondent may be asked a question on a specialised area, such as: 'Over the next five years or so, do you expect the use of value management to:

1 Increase a lot
2 Increase a little
3 No change
4 Decrease a little
5 Decrease a lot?'

Table 6.3 *Which of the following problems do you think need to be overcome in relation to the project management system?*

Please tick one or more of the following

No universally accepted definition of project management.	☐
No standard form of project management.	☐
Defining responsibilities on members of the project management team.	☐
Controlling the work packages of the project.	☐
Associated problem of communication.	☐
Higher cost associated with the project management system.	☐
Attitude of personnel involved in the project management project.	☐
Others (please state).	

The checklist format is a quick format but can be rather rigid. The response to each item does not have a degree of intensity, but a dichotomy. As an attribute, it is either applicable or not. Two examples of checklist questions are illustrated in this chapter. Table 6.1 was an exhibit of a checklist of questions taken from a dissertation topic on marketing related to Proposal 1 in Appendix 1. Table 6.3 is another exhibit where a checklist is used to find out the level of knowledge people have on a particular subject.

Grid

A grid is an elaboration of the checklist format, except it provides answers to two or more questions at the same time (see example in Table 6.4).

Rating scale

The rating scale is one of the most common formats for questioning respondents on their views or opinions of an object, event or attribute. Like the grid, rating scales can be regarded as an elaboration of the checklist format, except the respondent has the choice to express his/her degree of agreement or disagreement on a particular scale. The response categories of such questions are

Table 6.4 *Below are some well-known procurement methods that can be used to deliver projects according to client criteria. Please tick those procurement methods that you think are best to achieve each criteria listed*

Criteria	Procurement method		
	Traditional	Construction management	Design and build
Speed of total process Building complexity Building quality Cost of change Degree of price certainty Functionality Buildability			

called quantifiers: they reflect the intensity of the particular judgement involved. The following are some common sets of response categories:

1 Strongly agree (5 points) 1 Of great importance 1 More 1 Too little
2 Agree (4 points) 2 Of some importance 2 Same 2 About right
3 Neither/nor (3 points) 3 Of no importance 3 Less 3 Too much
4 Disagree (2 points) 4 Do not know
5 Strongly disagree (1 point)

1 Very satisfied 1 Very favourably 1 Very bad
2 Fairly satisfied 2 Favourable 2 Bad
3 Fairly dissatisfied 3 Not favourable 3 Would not matter
4 Very dissatisfied 4 Fairly good
5 Very good
6 It would depend

The numerical number shown beside the first category represents the intensity of the response. These are not usually shown in the questionnaire but are included here for your own use at the analysis stage.

Table 6.5 shows an extract from a questionnaire which was designed to investigate the views of the industry on factors affecting productivity on

Table 6.5   *Below are a number of factors which can have an impact on site productivity. From your experience, please express your opinion on how important each factor can be on influencing productivity. (Please tick the appropriate box)*

Factors influencing site productivity	Of great importance	Of some importance	Of no importance
1 Head office factors			
Delegation of responsibilities			
Integration of project information			
Project planning			
Scheduling of project activities			
Level of authority			
Supervision of subordinates			
Communication between head office and site			
Involvement of site managers in contract meetings			
Characteristics/attitude of site personnel			
The decision-making process			
2 Resource management effectiveness			
Procurement of materials			
Site programme			
Accuracy of the technical information			
Appropriateness of tools to be used for the tasks			
Knowledge of project technology			
Management interference on workmanship			
Site layout			
3 Motivational factors			
Promotion of employee			
Resentment regarding company policy			
Incentive schemes for good performance			
Opportunities to exercise skill/knowledge			
Management response to settle employee's grievances			
Workers uncertainty about career prospects			
Work environment			
4 Education and training			
Experience of employees			
Contract administration skill			
Knowledge of scientific techniques			
Training on new technology			
Availability of multiskilled project personnel			

construction sites. Respondents were asked to rate various factors they consider important. This type of rating is also called 'Matrix Rating'.

With the rating scales, it is your task to decide the number of scale points to use. Moser and Kalton (1993, p. 359) write:

> If the scale is divided too finely the respondents will be unable to place themselves within the scale, and if too coarsely the scale will not differentiate adequately between them. Often five to seven categories are employed, but sometimes the number is greater. The choice between an odd or even number depends on whether or not respondents are to be forced to decide the direction of their attitude; with an odd number there is a middle category representing a neutral position, but with an even number there is no middle category, so that respondents are forced to decide to which side of neutral they belong. Another factor to take into account in fixing the number of categories is that respondents generally avoid the two extreme positions, thus effectively reducing the number they choose between.

Likert scale

This type of scaling is similar to the rating scale, except the questions consist of attitudinal statements on the survey object (say attitude to job satisfaction) ranging from one extreme of favourableness to the other. Table 6.6 is an exhibit of a likert scale questionnaire extracted from Mullins's book, *Management and Organizational Behaviour* (1996).

Numerical rating scale

The numerical rating is another common questionnaire format. In this format, respondents are asked to give scores on an analogy, a service or an event. The rating may run out of 100, out of 10, out of 5 and so on. Examination marking is a typical example of a numerical rating scale. Table 6.7 on page 80 is an exhibit of a subjective way to tell the respondent's opinion on the criteria that are used in assessing project performance.

Ranking

The ranking format is used when the respondent is asked to place a set of attitudes or objects in ranking order indicating their importance, priorities or preferences. For example, in a survey on the level of job satisfaction,

Table 6.6 *The following statements are related to job satisfaction at work. Please indicate your reaction to each statement by ticking the appropriate cell*

Job satisfaction factors	Strongly agree	Agree	Neither agree nor disagree	Disagree	Strongly Disagree
1. Special wage increases should be given to employees who do their jobs well.					
2. Better job descriptions would be helpful so that employees will know exactly what is expected of them.					
3. Employees need to be reminded that their jobs are dependent on the company's ability to compete effectively.					
4. Supervisors should give a good deal of attention to the physical working conditions of their employees.					
5. Supervisors ought to work hard to develop a friendly atmosphere among their people.					
6. Individual recognition for above-standard performance means a lot to employees.					
7. Indifferent supervision can often bruise feelings.					
8. Employees want to feel that their real skills and capacities are put to use in their jobs.					
9. The company retirement benefits and share programmes are important factors in keeping employees in their jobs.					

10. Almost every job can be made more stimulating and challenging.								
11. Many employees want to give their best in everything they do.								
12. Management could show more interest in the employees by sponsoring social events after hours.								
13. Pride in one's work is actually an important reward.								
14. Employees want to be able to think of themselves as the best at their own job.								
15. The quality of the relationships in the informal work group is quite important.								
16. Individual incentive bonuses would improve the performance of employees.								
17. Visibility with upper management is important to employees.								
18. Employees generally like to schedule their own work and to make job-related decisions with a minimum of supervision.								
19. Job security is important to employees.								
20. Having good equipment to work with is important to employees.								

Source: Mullins (1996, pp. 514–16).

Table 6.7   *The following are a number of criteria which are commonly applied by clients of the building industry in assessing the performance of their projects. Please give a rating from 0 to 10 to each criterion according to the magnitude of importance each would be to your organisation (10 = very important)*

Criteria	Rating out of 10
1 Early start on site. 2 Reliability of the estimated design time. 3 Reliability of the estimated construction time. 4 Minimising the construction time. 5 Reliability of the estimated construction cost. 6 Obtaining a building at the cheapest cost. 7 Obtaining a building with low maintenance cost. 8 Obtaining a building with high aesthetic quality. 9 Obtaining a functional building to fit its purpose. 10 Input from the contractor related to buildability. 11 To have confidence in the contractor.	

respondents were asked to rank various dimensions they consider important in being in their jobs (see Table 6.8). There are, however, particular limitations to the number of rankings most people can be expected to carry out. Under normal survey conditions, to put ten objects in rank order is probably as much as can be asked (Oppenheim, 1996). Therefore, for a large number of objects you need to ask the respondents to rank for you a number of factors (say five factors) out of many factors as in the example shown in Table 6.9 on page 82.

Semantic differential scales

Diagrammatic rating scales is another means of measuring intensity of judgement and the semantic differential is the most popular form of this type of scaling. The technique was developed by Osgood to investigate the perceived meaning of various concepts. In this method, the respondent is asked to indicate his/her position on a seven-point bipolar scale defined with contrasting adjectives at each end. It can best be applied when the perceptions of different groups on a particular matter need to be compared. Examples could be good/bad, strong/weak, satisfied/dissatisfied.

Table 6.8 *Would you please indicate, in priority order, which of the follow. factors you find important in being satisfied with your job. Please indicate whicr. factor is first in importance, which is second, which is third and so forth (please circle the appropriate figure on the right-hand side)*

1	A well paid job (having a good income and being able to afford the good things in life)	1 2 3 4 5 6 7 8
2	A prosperous job (having a good opportunity to develop myself)	1 2 3 4 5 6 7 8
3	An important job (a job full of achievement that brings me respect and recognition)	1 2 3 4 5 6 7 8
4	A secure job (making certain that I have steady work and steady wages)	1 2 3 4 5 6 7 8
5	Relationship with fellow workers (getting along well with the people I work with)	1 2 3 4 5 6 7 8
6	Relationship with superiors (getting along well with my supervisor)	1 2 3 4 5 6 7 8
7	Job interest (good chance to do work of my interest)	1 2 3 4 5 6 7 8
8	Working condition (working within a good physical job condition)	1 2 3 4 5 6 7 8

The seven-point bipolar scale can be presented in numerical dimensions such as:

Good   —   —   —   —   —   —   —   Bad
      3   2   1   0 -1 -2 -3

or can be presented on a descriptive dimension like that shown in Table 6.10 on page 83. Numerical values can be assigned to the descriptive dimensions and comparisons can be made between items of various groups in terms of mean scores.

Criteria for constructing a questionnaire

The characteristics of a good questionnaire can be summarised as follows (Wood, 1991; cited in Naoum and Coles, 1997, pp. 18–20):

1 It must deal with a topic of some significance that is important enough to the respondent to merit a response.

Table 6.9 *In your view, what are the most serious factors causing accidents on construction sites? (Please select not more than five and number them 1, 2, 3, 4, 5 in order of seriousness)*

____ Age (very young or too old)	____ Lack of experience to execute the task
____ Lack of background safety training	____ Paying operatives danger money
____ Lack of personal care for safety	____ Poor enforcement of safety laws and regulations
____ Lack of safety training on site	____ Bad supervisor's safety behaviour
____ Bad workmates' safety behaviour	____ Bad subcontractors' safety behaviour
____ Mishandling of dangerous material	____ Improper fixing of scaffolding
____ Improper erection of steel	____ Lack of inspection for safety
____ Insufficient skill to operate plant	____ Lack of provision of safety clothing
____ Improper use of safety clothing	____ Improper use of safety equipment
____ Not issuing safety booklet	____ Lack of trade union involvement
____ Lack of site safety representative	____ Lack of safety policy
____ Lack of safety talk by management	____ Not displaying safety posters
____ Untidy site	____ Badly planned and disorganised site

2   It must seek information not obtainable from other sources. You should not ask people to do your data gathering for you especially when the data is readily available elsewhere.
3   It should be as short as possible but comprehensive enough to allow you to derive what you need without alienating the respondent.
4   It should be attractive in appearance, well laid out and well reproduced (see Appendix 3).
5   Where it contains directions they must be clear and complete.

Table 6.10 *Think of your leader and describe him/her by completing the following scale. Place an 'X' in one of the eight spaces according to how well the adjective fits your leader. Look at the words at both ends of the line before placing your 'X'*

	8	7	6	5	4	3	2	1	
Pleasant	–	–	–	–	–	–	–	–	Unpleasant
Friendly	–	–	–	–	–	–	–	–	Unfriendly
Rejecting	–	–	–	–	–	–	–	–	Accepting
Helpful	–	–	–	–	–	–	–	–	Frustrating
Unenthusiastic	–	–	–	–	–	–	–	–	Enthusiastic
Tense	–	–	–	–	–	–	–	–	Relaxed
Distant	–	–	–	–	–	–	–	–	Close
Cold	–	–	–	–	–	–	–	–	Warm
Co-operative	–	–	–	–	–	–	–	–	Uncooperative
Supportive	–	–	–	–	–	–	–	–	Hostile
Boring	–	–	–	–	–	–	–	–	Interesting
Quarrelsome	–	–	–	–	–	–	–	–	Harmonious
Self-assured	–	–	–	–	–	–	–	–	Hesitant
Efficient	–	–	–	–	–	–	–	–	Inefficient
Gloomy	–	–	–	–	–	–	–	–	Cheerful
Open	–	–	–	–	–	–	–	–	Guarded

6 Unless there is a very good reason for another format, questions should be arranged in categories which allow easy and accurate responses.

7 Questions must be as objective as possible without offering leading questions. For example, a question such as 'Do you not agree that site managers should have the right to have a say in finalising the building design?' Here, it might be difficult for site managers to answer 'No' in response to that question.

8 In their sequencing, questions should run from the general to the specific, from simple to complex, and from those that will create a favourable impression upon the respondent to those that may be sensitive.

9 Bearing in mind point (8) above, you should avoid questions that may annoy or embarrass the respondent.

10 The questionnaire must provide for ease of tabulation and/or interpretation, and should be designed accordingly.

The above points illustrate the importance and the role of the questionnaire to the survey's success. The design of the questionnaire also requires careful attention, and consideration should be given to:

1 Examining other questionnaires to identify possible clues.
2 Obtaining as much help as possible in order to make sure it is easy to be understood by different audiences.
3 Allowing for a pilot stage at the draft phase in order to get valuable responses and to detect areas of possible shortcomings.
4 Close liaison with the supervisor and make sure that he/she has seen and approved the questionnaire.
5 Trying different question orders at the pilot stage. This may assist in finding the best approach.
6 Experimenting with different types of questions, both open and closed.
7 In order to obtain the maximum number of respondents and collect the most relevant information make sure:
   (a) You identify the sender and the person who receives the returned questionnaires.
   (b) You know the purpose of the research and the questionnaire.
   (c) There is an incentive for the respondent to complete the questionnaire.
   (d) You reassure the respondent of the confidentiality of the questionnaire.
   (e) You allow the respondent the opportunity to provide some personal insight not contained elsewhere.

Note to students: When trying to design the questionnaire, you need to ask yourself a number of questions such as:

1 Are you using quantitative or qualitative measures?
2 Is your study a comparative or a singular investigation?
3 If it is a comparative study, how many survey groups are you investigating and who are they?
4 How do the groups differ in their characteristics?
5 What is the depth of the investigation? Is it a specific problem or a generalized one?
6 How many responses or cases do you need in order to answer your questions and achieve your objectives?
7 If you are testing a theory, are you using control variables and why?

Once you have answered the above questions and designed the questionnaire, you can then invite the respondents to fill out the questionnaire or rate them or do whatever is required. It is absolutely essential to do a complete pilot study before you collect the final data for your main study.

A pilot study

Whenever you construct a questionnaire, it is advisable to complete a pilot study before you collect the final data from the whole sample. A pilot study provides a trial run for the questionnaire, which involves testing the wording of the question, identifying ambiguous questions, testing the technique that you use to collect the data, measuring the effectiveness of your standard invitation to respondents, etc. Bell (1996, p. 84) described a pilot study as: 'getting the bugs out of the instrument (questionnaire) so that subjects in your main study will experience no difficulties in completing it and so that you can carry out a preliminary analysis to see whether the wording and format of questions will present any difficulties when the main data are analysed.' Bell went further and noted, ask your guinea pigs the following questions:

1 How long did it take you to complete?
2 Were the instructions clear?
3 Were any of the questions unclear or ambiguous? If so, will you say which and why?
4 Did you object to answering any of the questions?
5 In your opinion, has any major topic been omitted?
6 Was the layout of the questionnaire clear/attractive?
7 Any comments?

You will probably be amazed by how much you will learn from a pilot study. The little extra time it takes to complete a pilot study test will be well spent.

Covering letter

After constructing your questionnaire, your next step is to write a covering letter to accompany your questionnaire. The letter should explain the purpose of the survey or questionnaire in order to encourage a high response. The content of the covering letter is particularly important in postal questionnaires because it is the only way you can persuade the subjects to respond to your questionnaire. Nachmias and Nachmias (1996) write:

A cover letter must succeed in overcoming any resistance or prejudice the respondent may have against the survey. It should (1) identify the sponsoring organization or the persons conducting the study, (2) explain the purpose of the study, (3) tell why it is important that the respondent answer the questionnaire, and (4) assure the respondent that the information provided will be held in strict confidence.

Dear Sir/Madam

Re: Dissertation – BSc Construction Management

I am currently undertaking a Bachelor of Science Degree in Construction Management at London South Bank University. In fulfilment of this dissertation I am required to research a topic area and produce a dissertation. The topic I have chosen is 'Project Management as a Method of Building Procurement' and I am investigating the following aspects of its use:

1   The criteria which clients feel important in the selection of Project Management contracts.
2   The types of building for which Project Management is used and the dominant reasons for the selection of this type of procurement path for these buildings.
3   The types of clients who are predisposed to use Project Management.

I would be very grateful if you could complete the attached questionnaire and return it in the addressed, prepaid envelope provided. Needless to say, the information provided will be treated with strict confidence and individual firms will not be identified. Equally, a copy of the summary report will be available to the co-operating firms.

Yours faithfully,   Supervisor

Signature   Signature

Figure 6.3   *Example of postal survey covering letter*

An example of a covering letter that can be used with a postal survey is presented in Figure 6.3.

Summary

This chapter described the stages for constructing a questionnaire together with types and question formats. The foundation of all questionnaires, whether mailed or to be used for interviewing, is the questions. The principle that you need to follow in constructing your questionnaire is to go back to your dissertation proposal and start to ask specific questions related to the aim, objectives and hypothesis of your study.

Questions can be of two types: open ended and closed ended. Open-ended questions have the advantage of giving an opportunity to respondents to express their views on the subject but can be rather difficult to analyse the results later. By contrast, closed-ended questions require a short response and their analysis is straightforward. However, you need to have sufficient knowledge on the subject of your investigation in order to offer respondents a set of response categories from which they should choose the one that most represents their opinions, views, attitudes or perceptions.

Questions can be either factual or an opinion type. Factual questions are designed to obtain objective data while opinion questions are designed to elicit subjective information. The questions can be formulated in various ways. The most common formats are checklist, grid, rating scales, numerical rating scales, ranking and semantic scales. These formats are described in this chapter with examples. It is advisable to conduct a pilot study to test the wording of the questions, check the length of the questionnaire and make sure that the questions are not ambiguous. Finally, you need to write a covering letter to accompany your questionnaire stating who you are, what is the purpose of your study, ensuring confidentiality and perhaps offering them an incentive to respond.

References

Bell, J. (1996) *Doing Your Research Project: A Guide for First Time Researchers in Education and Social Science*. Open University Press.

Lazarsfeld, P. (1944) The Controversy over Detailed Interviews: An Offer for Negotiation. *Public Opinion Quarterly*, **8**: 38–60.

Moser, C. and Kalton, G. (1993) *Survey Methods in Social Investigation*. Dartmouth. p. 359.

Mullins, L. (1996) *Management and Organizational Behaviour*. Pitman.

Nachmias, C. and Nachmias, D. (1996) *Research Methods in the Social Sciences*, fifth edn. Arnold.

Naoum, S. and Coles, D. (1997). *Dissertation Guide: Instructions for preparation, control and presentation of the dissertation*. School of Construction, London South Bank University.

Oppenheim, A. (1996) *Questionnaire Design, Interviewing and Attitude Measurement*. Pinter. p. 112.

Part Three

Analysis and Presentation of the Results

7 *Measurements and probability*

Measurement is a procedure in which a researcher assigns numerals (numbers or other symbols) to empirical properties (variables) according to rules. It is closely linked to the research approach and questionnaire construction which were discussed in Chapters 4 and 6 respectively. There are four levels of measurement: nominal, ordinal, interval and ratio. In some cases your research involves a search for a measure that is already developed and in other cases you need to design a measure. This chapter focuses on these levels of measurements in order to prepare the ground for the next exciting chapter: analysis of the results. This chapter also provides a statement of probability which is an important term to understand in testing your research hypothesis. The contents of Chapter 7 are illustrated in Figure 7.1.

Note to students: The term 'measurement' is mainly used to quantify quantitative questions. I prefer to use the terms 'evaluation' or 'assessment' to process exploratory questions. Analysis of both quantitative and exploratory questions is discussed in Chapter 8.

Level of measurement

In order to be able to select the appropriate method of analysis, you need to understand the level of measurement. For each type of measurement, there is/are an appropriate method/s that can be applied and not others.

Measurement is a procedure in which a researcher assigns numerals (numbers or other symbols) to empirical properties (variables) according to rules. Suppose you invited five contractors (A, B, C, D and E) to tender for building your project. Having found that the difference in price among five tenderers is minute, you decided to make the selection on the basis of which contractor best meets the following criteria: speed, quality and safety. These three criteria vary. For example, one contractor may deliver the project earlier and have a safety programme, but the quality of the finished product may be unsatisfactory. Accordingly, you decide to rank each of the three criteria on a scale of five numbers: 1, 2, 3, 4 and 5. Number 1 indicates total dissatisfaction, and number 5 stands for complete satisfaction. You then evaluate the five tenderers.

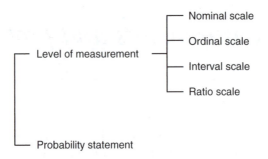

Figure 7.1   *Contents of Chapter 7*

After examining the scores, you may decide to appoint contractor (B) because it received the highest score on the three counts. The numbering process discussed above is a crude example of explaining the nature of measurement. However, it conveys the basic idea expressed in the definition of measurement, i.e. you assigned numerals to properties according to rules. The following section will explain the four principle levels of measurement, namely, nominal, ordinal, interval and ratio. Your primary data collection should fall within one or more of these levels.

Nominal scale

Nominal numbering implies belonging to a classification or having a particular property and a label. It does not imply any idea of rank or priority. Nominal numbering is also conventional integers, that is positive and whole numbers (this may well be due to the fact that most statistics are analysed by computer which, as you know, handles numbers more easily than letters or strings). For example, if you conducted a survey to investigate the use of a particular project management software in the construction industry, you will count the number of companies using the package and categorise them as shown in Table 7.1.

The numbers given to the categories in Table 7.1 are nominal, i.e. name only. For example, number 1 (contractors) is not half of number 2 (architects) or in

Table 7.1   *Categorisation table of companies using project management software*

Category	1	2	3	4	5
	Contractors	Architects	Q.S.	Engineers	Others
	20	70	35	40	10

any way prior to or less than the others in quantity. The numbers are simply convenient but arbitrary labels for identifying each type of company. You could have used the label (A) instead of 1, (B) instead of 2, (C) instead of 3 and so on. The numbers within each category are called frequencies. The frequency distribution and the terms associated with it will be discussed in Chapter 8.

Ordinal scale

This is a ranking or a rating data which normally uses integers in ascending or descending order. An example of an ordinal is when you ask an attitudinal question. For instance, you may want to measure your respondents' attitude to a motivational statement such as 'individual incentive bonuses would motivate people to work harder and in turn improve the performance of employees'. Here, the respondent is asked to mark a number representing his/her degree of agreement or disagreement with this statement. Table 7.2 illustrates a possible coding frame between the numbers and the answers. The numbers assigned to the agreement scale (5, 4, 3, 2, 1) do not indicate that the interval between the scales are equal, nor do they indicate absolute quantities. They are merely numerical labels.

Another example of an ordinal scale is when respondents are asked to rank items by their own preference. For instance, if we asked eight people to rank the quality of a particular product in order of their preference, we might obtain Table 7.3.

Similar to the previous example, the numbers in Table 7.3 do not indicate that the interval between the ranks are equal. Although the numbers are equally spaced, it does not imply that the property each represents is also equally spaced. If two respondents have the rank 8 and 6 and two others are ranked 7 and 5, it does not mean that the differences between the two pairs are equal.

Table 7.2 *Ordinal ranking scale*

	Strongly agree	Agree	Neutral	Disagree	Strongly disagree
	5	4	3	2	1
Site managers	4	3	5	5	0
Head office personnel	0	1	1	9	6
Operatives	7	5	4	1	0

Table 7.3   *Quality of product scores*

Person	Score	Rank of score
1	13	5.5
2	20	7
3	9	1
4	13	5.5
5	10	3
6	10	3
7	10	3
8	24	8

Notice the tied ranks in Table 7.3. In this example the score of 9 was given the first rank. As three people share the score of 10, you need to share the ranking, such as:

$$\frac{2 + 3 + 4}{3} = \frac{9}{3} = 3$$

For subject one and four this means the rank 5 and 6 are shared, i.e.:

$$\frac{5 + 6}{2} = \frac{11}{2} = 5.5$$

Interval scale

The numbering system in the ordinal and nominal measurement is purely an arbitrary label for identifying each type of person. If you have a set of obser-vations or data where the distance between each observation is constant, then this type of measurement is called an interval level of measurement. Often used examples are minutes, kilograms, number of words recalled in a memory test or percentage marks in the exam. The interval between 20 to 30 minutes is the same as 50 to 60 minutes. When it comes to numerical scores, such as num-bers of items recalled per minute, you are dealing with numbers and you can assume that the distances between scores are the same. This type of measure-ment is another example of interval measurement because it assumes equal intervals between the data on a continuous scale.

Ratio scale

The ratio scale is similar to the interval scale except it involves the kind of numerical scale which has a natural zero such as age, salary, time and distance. However, you do not need to bother about the difference between interval and ratio scales. For the level of statistics described in this book, both measurements are treated in exactly the same way.

Probability statement

The subject of probability is an important term to understand when you start to analyse your results. Statistical probability will tell you whether any differences in scores are due to your manipulation of the variables, as predicted by your research hypothesis or, alternatively, whether the differences are only due to chance fluctuations as stated in the null hypothesis. You do not need to be bothered now about the term 'null hypothesis', it is discussed later in Chapter 8.

We use the word 'probably' almost everyday to express our views on certain things. Consider the following statements:

1   It will probably rain next week.
2   I will most probably visit my friend tomorrow.
3   I will definitely pass my maths exam.

In the first statement you are giving a certain degree of chance that it will rain tomorrow. This chance is normally measured as a percentage out of 100. It could be 1 per cent, 5 per cent, 20 per cent, 50 per cent and so on. In the second statement you are making a more assured statement by saying 'most probably'. Again this could be any percentage out of 100, but rather towards the upper limit of the scale. It could be an 80 per cent, 90 per cent, 95 per cent chance and so on, but not 100 per cent. In the third statement you are making a more certain statement by using the word 'definitely' (i.e. 100 per cent). But on what basis have you made these statements or judgements. The second statement may be based on your experience, but the first and third statements are based on historical records. For example, your past record shows that you never failed the maths test in the past, hence, you are eliminating the chance of failing this time.

The term 'probability' can, therefore, be defined as the percentage that an event occurs in a number of times. In observational or experimental studies, if 1000 tosses of a coin result in 529 heads, the relative frequency of heads is

529/1000 = 0.529. If another 1000 tosses result in 493 heads, the relative frequency in the total of 2000 tosses is (529 + 493)/2000 = 0.511. According to the statistical definitions, by continuing in this manner we should ultimately get closer and closer to a number which we call the probability of a head in a single toss of the coin. From results so far presented this should be 0.5 to one significant figure. To obtain more significant figures, further observations/experiments must be made.

Similarly, scientific and attitudinal research, related to the built environment, can be based on primarily probabilistic or indicative explanation. For example, a researcher might state that traditional contracts will most probably overrun on cost. His/her rationale is that, from past project records, in 8 out of 10 occasions the project exceeded the budget when using the traditional form of contract. But how significant is your result, i.e. how confident or sure are you of arriving at this concluding remark? The answer to this is, you have to apply statistical tests to determine the direction of your research. Chapter 8 provides details of the most popular statistical tests for the level of students reading this book.

Note to students: The statistical tests discussed in Chapter 8 will provide you with a probability that allows you to judge whether your results are significant or are due to chance. Later, you will find out that all the given examples will end up by stating the probability figure of the statistical test. For instance, you will come across an expression $P < 0.05$. This means that the probability of a result being due to chance is less than 5 per cent or 5 in 100. Another, $P < 0.01$, means that the probability of a result being due to chance is less than 1 per cent or 1 in 100, and so on. Therefore, the less the probability figure the more confident you can be in concluding that there is a significant difference applied to your data.

The conventional level by which you can reject the null hypothesis and conclude that the results are significant is, $P < 0.05$. Otherwise, the results of your test are not significant and you have to accept the null hypothesis of no relationship or association between the research variables.

Summary

This chapter discussed the level of data measurements and explained the term 'probability'. Measurement is a procedure in which you assign numerals (numbers or other symbols) to empirical properties (i.e. variables) according to rules. There are four levels of measurement and the data that you will collect should fall within one or more of these levels. The four levels are known as: nominal; ordinal; interval and ratio. The nominal level implies belonging to a classification or having a label and does not imply any rank or priority. The ordinal level refers to data that are ranked or rated in ascending or descending order. The interval level is when you have a set of observations where the distance between each observation is constant, e.g. minutes, kilograms and exam texts. The ratio level is similar to the interval except it involves a natural zero.

The term 'probability' is defined as the percentage that an event occurs in a number of times. It is calculated to determine the direction of your study. The statistical tests that are explained in Chapter 8 will allow you to establish the confidence level of your research by testing the hypothesis. The probability figure, that is calculated using the statistical tests, will give you a level of significance by which you can judge whether to approve or disapprove your research hypothesis.

Additional reading

Green, J. and D'Oliveirs, M. (1993) *Learning to Use Statistical Tests in Psychology: A Student's Guide.* Open University Press.

Nachmias, C. and Nachmias, D. (1996) *Research Methods in the Social Sciences*, fifth edn. Arnold.

Swinscow, T. (1997) *Statistics at Square One.* British Medical Association.

8 *Analysis of the results*

Once you have collected the data, you will be ready to analyse the results to determine the direction of the study. You will be gathering a lot of information which makes it difficult to present every bit of it. Therefore, it is expected that you give a summary of the data which highlights main trends and differences in the most appropriate manner. At this stage of your research study you will ask yourself a number of questions. For example, will I use a frequency distribution table or a bar chart to analyse each question in the questionnaire? Will I use a total score or should I analyse the data separately for each item in the question-naire or rating scale? Which is the most appropriate test to use with these data, the t-test or the chi-square test? Should I compute some rank order correlation? Once you and your supervisor are satisfied with the answers to these questions, you can apply the statistics to find out if there were significant results.

This chapter describes, with examples, the methods of analysis that are commonly used to summarise and organise the data in a most effective and meaningful way. The chapter will cover the following:

1 Exploratory data analysis (open-ended questions).
2 The descriptive method of analysis including: measurement of central ten-dency, the normal curve and the frequency distribution.
3 The inferential statistical method of analysis including: the 't-test', chi-square test, Spearman 'rho' ranking correlation and the product moment correlation coefficient.

The contents of Chapter 8 are illustrated in Figure 8.1.

Exploratory data analysis (open-ended questions)

As discussed in Chapter 4, exploratory research is a qualitative research and can be described as a social or organisational behaviour research which produces

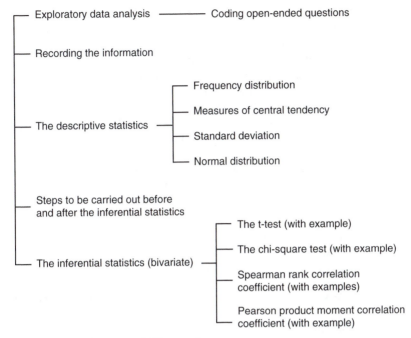

Figure 8.1 *Contents of Chapter 8*

results that are not obtained by statistical procedures or other methods of quantification. This type of data analysis refers to research about people's lives, their stories and behaviour, and it can also be used to examine organisations, relationships and social movements. Some of the data may be quantified afterwards, but the analysis is qualitative.

Research carried out in this way produces descriptive data such as people's own spoken or written words or observable behaviour. (Note that observational and experimental studies are not covered in this book.) The instrument or tool that is often used to collect exploratory research data is the open-ended type of questionnaire. The following section explains the procedure for processing open-ended questions.

Coding open-ended questions

Open-ended questions can be used in postal questionnaires as well as in interviews. The usual reason for using open-ended questions is that the researcher has no clear hypotheses regarding answers, which will be numerous and varied. Analysis of the open-ended questions can be rather complicated and not as straightforward as structured closed-ended questionnaires. It also requires a great skill to accurately report the information.

The best way to analyse open-ended questions is to code the information in terms of ideas and themes. The purpose of coding such questions is to reduce the large number of individual responses to a few general categories of answers that can be assigned a numerical code. In order to analyse open-ended questionnaires the following steps can be taken:

1   Place all similar answers in a general category and assign them a code. For example, individual clients asked about the use of construction management for their projects might give the following answers:

    (a) We build large and complex projects.
    (b) We want to have good control on the project.
    (c) It is quicker.
    (d) Construction management provides high integration.

    All of these answers can be categorised under 'favourable to construction management'.
    Other clients might give the following answers:

    (a) We build housing projects.
    (b) We are not sophisticated clients.
    (c) It is too risky.
    (d) It is not a straightforward process as with the JCT contract.

    All of these answers can be categorised under 'unfavourable to construction management'.

2   After establishing the general categories, you then need to divide them into, say, 2–4 sub-categories, and assign them a code. Coding is the process of identifying and classifying each answer with a numerical score or other character symbol. It usually involves entering the data for analysis or for computer storage. The coding categories should be exhaustive and provide for all possible responses. They should be mutually exclusive and independent so that there is no overlap among categories. On highly structured questionnaires, the categories may be precoded. With open-ended questions, the answers are postcoded. This means that the categories are assigned after the data have been collected. The categories must be assigned according to your judgement. It is better to assign too many categories than too few because it is easier to collapse several categories into one than to increase the number of categories (Zikmund, 1997).

Table 8.1 shows an example of constructing categories and subcategories to an open-ended question, and devising their codings scheme. These categories are for clients' responses to a question that asked for comments about the use of construction management for their projects.

3 After establishing the coding, choose the method of analysis which you think is suitable for your data (descriptive or/and inferential method). These methods are described later in this chapter. The suitability of each method depends on the nature of data, the type of measurement and your sample size.

Recording the information

Recording the information is an important step in the research process. Before starting any form of analysis you need to devise a data summary sheet for your data. The actual process of transferring the data from the questionnaire or data collection form (after the data have been collected) into a data summary form is called the 'recording scheme' or 'production coding'. Table 8.2 is an example of a data summary form.

The first column of the data summary form is usually designated for the respondent's number. The rest of the columns represent the coded answers to the questions in the questionnaire. For example, question 1 in Table 8.2 may be an opinion question requiring the respondent to indicate strong agreement (coded 3), neither agreement nor disagreement (coded 2) or strong disagreement (coded 1). Question 2 may require a response of 'Yes' (coded 1) or 'No' (coded 2) and so on.

Note to students:

1 The principle of the data summary form can be applied to open-ended questions (postcoded) as well as to closed-ended questions (precoded)
2 The data summary form is usually inserted in the appendix of your dissertation because they are known as 'raw data'.

After recording the information, the field of statistics can be applied to the raw data so that some kind of interpretation and discussion can be made on the results. There are two methods you can use to analyse your data: the descriptive statistics method and the inferential statistics method (also known as bivariate statistical analysis). Both are explained below.

Table 8.1   *Example of coding open-ended questions*

Favourable to construction management

Sub-category 1 – Team relationship
Code
10   Smooth running project
11   Right attitude of team members
12   Contract of trust
13   High integration
14   Good communication
15
. . .
19   Other team relationship

Sub-category 2 – Project control
Code
20   High control on subcontractors
21   High control on safety
22   Good control on variations
23   Good control on quality
24
25
. . .
29   Other project control

Sub-category 3 – Project outcome
Code
31   Would not change the project objectives
32   High certainty on programme time
33   High certainty on project cost
34   Safer project
36
. . .
30   Other project outcome

Sub-category 4 – Miscellaneous
Code
41   Good for complex projects
42   Architects recommended
43   Concentrates on client needs
44   Would not cut corners
46
. . .
49   Other miscellaneous

Unfavourable to construction management

Sub-category 5 – Team relationship
Code
50   Too complicated
51   Does not eliminate 'them and us' attitude
52   Conflict with the professionals
53
54
. . .
59   Other team relationship

Sub-category 6 – Project control
Code
61   Difficult to control the programme
62   Does not control subcontractors
63   Difficult to control supervisors
64
65
. . .
69   Other project control

Sub-category 7 – Project outcome
Code
71   Expensive method
72   Not economical for small projects
73   Too risky to clients
74
75
. . .
79   Other project outcome

Sub-category 8 – Miscellaneous
Code
81   Do not like the concept
82   Use what our architect recommends
83   No universal accepted contract
84   The concept comes in different formats
85   Difficult to define and agree the work packages
. . .
89   Other miscellaneous

Table 8.2   *Example of a data summary form*

Respondent Number	Question 1 Code number 1 2 3	Question 2 Code number 1 2	Question 3 1 2 3	Question 4 ... 1 2 3	Question 30 1 2 3
1.	3	1	1	2	3
2.	3	2	3	3	3
3.	2	2	1	3	3
4.	2	2	3	3	3
5.	3	2	3	3	3
6.	3	2	3	3	3
. . .					
37.	1	2	2	1	2

The descriptive statistics method

The descriptive statistics method is the simplest method of analysis which provides a general overview of the results. It gives an idea of what is happening. The descriptive method will either analyse the responses in percentages (as in the case of a large sample) or will contain actual numbers (as in the case of a small sample). In this section I shall introduce three formal terms which are used to describe aspects of a group of data; these are 'frequency distribution', 'measurement of central tendency' and 'measurement of dispersion'.

Frequency distribution

When summarising large amounts of raw data it is often useful to distribute the data into categories or classes and to determine the number of individuals or cases belonging to each category. This is called 'category frequency'. It can be presented in the form of tabulation, a bar chart, a pie chart or a graph.

Tabulation

Tabulation is the simplest way to show the frequency of observations of each response to each variable under investigation. To conduct a frequency distribution you simply list the categories of the variable and count the number of observations in each. Table 8.3 shows the frequency distribution of the results related to 'an investigation of the marketing philosophy of private house-builders'. In this example, out of the 37 house-builders, 8 respondents fall into category A and 29

Table 8.3   *Frequency distribution of marketing philosophy*

Question	No. of respondents in Category A – 'We first build the estate and then work out how best to sell the product'	%	No. of respondents in Category B – 'We first determine customers' wants and then design the product'	%
In the house-building cycle, which action best describes your firm's business philosophy?	8	20	29	80

respondents fall into category B. The frequencies can then be converted into percentages of the total number. Therefore, about 20 per cent of the house-builders believed that their firm's business philosophy is to 'first build the estate and then work out how best to sell the product' (i.e. category A), while 80 per cent believed that they 'first determine customers' wants and then design the product types and packages to satisfy those wants' (i.e. category B).

The other type of frequency is when two or more different categories of people respond to the questionnaire and the sample is small. For example, Table 8.4 shows results of clients' satisfaction on the performance of time of construction management contracts.

Table 8.4   *Level of clients' satisfaction on the performance of time of construction management contracts*

	High	Low	Total
Public clients (N = 5)	4	1	5
Private clients (N = 10)	7	3	10
Total	11	4	15

N = number of sample

Table 8.5 *Frequency distribution of injuries to employees in the construction industry for the year 1994/95 (provisional), based on secondary data gathered by the HSE's Field Operations Division (see raw data in Appendix 5)*

Worker occupation	No. of major accidents	%	No. of minor accidents	%
Bricklayer	72	4.0	550	5.9
Carpenter/joiner	136	7.5	1127	12.1
Electrician	108	6.0	330	3.5
Ground worker	58	3.2	139	1.5
Pavior	39	2.2	443	4.7
Scaffolder	62	3.5	222	2.4
Road worker	64	3.6	195	2.1
Driver	120	6.7	539	5.7
Labourer	316	17.6	1659	17.7
Maintenance	60	3.3	294	3.1
Manual production	104	5.8	548	5.8
Other	659	36.6	3329	35.5
Total	1798	100.0	9375	100.0

Notice that Table 8.4 shows only the actual frequencies with no percentages. It is important to use actual numbers when the sample is small as in the case of limited interviews. This is because a statement on the above figures, such as '80 per cent of public clients were highly satisfied with the time of construction management', is factually accurate but can be misleading as the reader will assume a much larger total number of public clients than five.

Table 8.5 is an example of a frequency distribution applied for analysing secondary data collection. In this particular example, the data are the frequency of accidents record, published by the HSE, of 12 occupations.

Note to students: It is essential that, after presenting the results, you need to make comments on and interpretations of your findings. What does this finding tell you? What does it mean? Does it correspond closely with previous studies or not. Chapter 9 provides you with a guidance on how to comment and interpret your results.

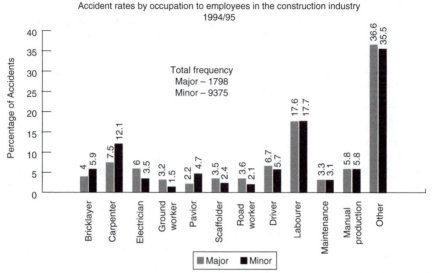

Source: Data taken form HSE Statistical Services Unit – Liverpool, 1996

Figure 8.2   *Example of a histogram presentation*

Bar charts/histograms

Horizontal or vertical bar charts can be used with the frequency distribution table or as an alternative method of presentation in the dissertation. Figure 8.2 is a bar chart for the data presented in Table 8.5. The variable (worker occupation) is placed on the horizontal axis and the percentage frequency on the vertical axis. The actual frequencies are included at the top of the bars and the total frequency of $N = 1798$ for major accidents and $N = 9375$ for minor accidents is stated in the conventional format. The height of the bars indicates the proportion of the frequency.

Pie charts

Another method of presentation that is widely used is the pie chart. The areas of the pie chart represent the proportion of the respondents and are usually presented as percentages. Figure 8.3 is a pie chart representing the percentages of alternative contractual arrangements.

Note to students: The method of presentation chosen (tabulation, compound bar chart, component bar chart, graph or pie chart) depends on your judgement as to which presentation is the clearest. Remember, always consult your supervisor.

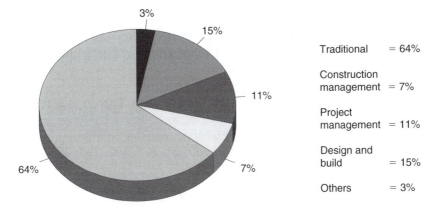

Figure 8.3   *Pie chart example representing the proportions of alternative contractual arrangements in the UK in 1996*

Measurement of central tendency

This type of analysis is applied when you have a group of data and you wish to find the most typical value for the group, or the score which all other scores are evenly clustered around. These statistics are known as the 'mean', the 'median' and the 'mode'.

The mean

This is the average of all the values in a set of data. The mean is calculated by adding all the values in the group and then dividing by the number of values. Hence, the arithmetic mean of the cost per square metre of five contracts £564, £505, £556, £445 and £530, is taken as:

$$\frac{564 + 505 + 556 + 445 + 530}{5} = \frac{2600}{5} = £520$$

The median

The median of a set of numbers arranged in order of magnitude (i.e. in an array) is the middle value or the arithmetic mean of the two middle values. To establish the median you need to arrange the set of data in an array. Hence, the array of the above set of numbers is 445, 505, 530, 556, 564 and the median is, hence, 530.

The mode

The mode of a set of numbers is that value which occurs with the greatest frequency, i.e. the most common value. The mode may not exist, and if it does exist it may not be unique. For example:

1   The set 220, 220, 500, 700, 900, 900, 900, 1000, 1000, 1100, 1200 has mode 900.
2   The set 300, 500, 800, 1000, 1200, 1500, 1600 has no mode.
3   The set 200, 300, 400, 400, 400, 500, 500, 700, 700, 700, 900 has two modes, 400 and 700.

Measurement of dispersion based on the mean

This type of analysis can show you the degree by which numerical data tend to spread about an average value and it is called 'variation' or 'dispersion'. It is represented by the formula:

$$\text{Mean deviation} = \frac{\Sigma(X - \bar{X})}{N}$$

Hence, if the cost per square metre of five projects was taken and the results were £505.00, £557.00, £465.00, £458.00 and £530.00, the mean deviation can be calculated as follows:

$$\text{Arithmetic mean} = £503.00$$

$$\text{Mean deviation} = \frac{\begin{aligned}|505 - 503| + |557 - 503| + |465 - 503| + \\ |458 - 503| + |530 - 503|\end{aligned}}{5}$$

$$= \frac{141}{5} = 28.2$$

Standard deviation

The standard deviation is another measure of the degree in which the data is spread around the mean. The theoretical basis of the standard deviation is complex and need not trouble the ordinary user (most scientific calculators have a function to calculate the standard deviation). But a practical point to note is that the data should fall approximately into a so-called 'normal distribution'. When it does so the standard deviation provides a useful basis for interpreting the data in terms of probability (see Probability statement in Chapter 7).

The formula for the standard deviation (SD) is

$$SD = \sqrt{\frac{\Sigma (X - \bar{X})^2}{N}}$$

Hence, the SD for the five unit cost = £35.70.

Normal distribution

Normal distribution is an important expression in the field of statistics because the selection of some inferential statistics depends on whether the data is normally distributed or not. The normal distribution curve can tell you a great deal about the scatter or distribution of the data (see Figure 8.4). The principle properties of the normal distribution curve are as follows (Nachmias and Nachmias, 1996):

1 It is symmetrical and bell-shaped
2 The mode, the median and the mean coincide at the centre of the distribution.
3 The curve is based on an infinite number of observations.
4 A single mathematical formula describes how frequencies are related to the value of the variable.
5 In any normal distribution, a fixed proportion of the observations lies between the mean and fixed units of the standard deviation. A range covered by one standard deviation above the mean and one deviation below it ($\bar{X} \pm 1SD$) includes about 68 per cent of the observations. A range of 2 standard deviations above and 2 below ($\bar{X} \pm 2SD$) includes about 95 per cent of the observations. And a range of 3 standard deviations above and 3 below ($\bar{X} \pm 3SD$) includes about 99.73 per cent of the observations. Consequently if you know the mean and standard deviation of a set of observations, you can obtain some useful information by simple arithmetic. By putting 1, 2 or 3 standard deviations above and below the mean you can estimate the ranges that would be expected to include about 68 per cent, 95 per cent and 99.7 per cent of the observations.

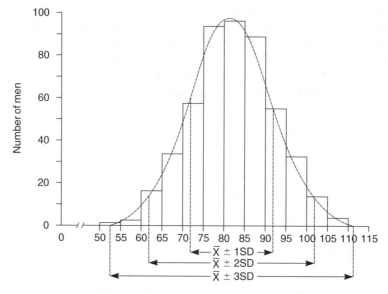

Figure 8.4   *A typical normal distribution curve*

The inferential statistics method (also known as bivariate statistical analysis)

Many survey analyses entail the comparison of results for different parts of the sample, for example the proportion of traditional contracts that overrun on time with the proportion of design and build contracts that overrun on time. In this case one might carry out a statistical significance test of the difference between the proportions. Statistical tests vary in their applications. Some tests are appropriate for testing differences in scores, other tests are suitable for assessing whether two sets of scores are correlated. For example, if you have two samples of normally distributed data and you want to find out whether there is a significant difference in the mean between the two samples, then you choose the t-test. If you have two or more categories of people and you want to find out whether there is significant difference in their attitude towards certain issues, then you use the chi-square test. If two groups of people are ranking a range of criteria and you want to know whether the differences in their ranking are significant, you will apply a ranking correlation (rho) and so on. The following sections will describe these tests in detail. Other statistical tests are also available but they are left for you to review. But before describing these tests, there are five steps that you need to consider.

Steps to be carried out before and after the inferential (bivariate) statistical tests

Step 1: Formulate your research hypothesis in terms of the predicted results (like the one stated in your proposal – see examples in Appendix 1)

Step 2: State the null hypothesis

The null hypothesis is a statement that you make which is the antithesis of your research hypotheses. It is usually an expression of no difference or no relationship between the variables. Suppose that the research hypothesis (Hi) states: 'Projects delivered under the traditional contracts are more expensive than similar projects delivered under the design and build contracts.' With the mean score for the cost of projects delivered under the traditional contracts designated as X1 and in the design and build contracts as X2, the research hypothesis, symbolised by (Hi) will be:

Hi: $X1 > X2$

The null hypothesis, symbolised by (Ho) will be:

Ho: $X1 = X2$

This means that your null hypothesis states: 'There is no significant difference in the mean cost between traditional contracts and the design and build contracts.' After applying the appropriate test you will find out whether your null hypothesis stands true or false.

Step 3: Decide which test to use

Choose the appropriate statistical test for your data to determine the direction of your investigation. There are two main kinds of statistical tests that you can apply, known as 'parametric' and 'non-parametric' tests. A parametric test is a statistical test based on several assumptions about the parameters of the population from which the sample was drawn. To apply a parametric test, your data must comply with certain conditions. Among the most important conditions are the assumptions that the observations must be drawn from a normally distributed population and be of an interval scale (see discussion of normal distribution above) whereas a non-parametric test does not specify the normality condition. There are certain assumptions associated with most non-parametric tests, however, they are weaker than those associated with parametric tests. I will introduce the rationale, the rules and the steps to apply for each test as I go along.

Deciding the significance of your results also include tests for correlation. These kinds of tests measure the strength of the relationship between two or more variables and are usually applied on analytical studies. I will introduce the 'Spearman rank correlation' and 'Pearson product moment correlation', as they are the most common tests and suitable for the level of students reading this book.

Step 4: Calculate and obtain the test statistics

After you decide on the test, you need to carry out the appropriate calculations on your research data. There are two ways in which you can carry out the calculations:

1   Manually.
2   Using a software package such as SPSS, Minitab and Statsmaster. These packages are not described in this book as they are beyond the scope of its contents. This is not to say that these packages are complex or difficult to understand. On the contrary, they are easy to learn and once you get the hang of using them, you will find it very easy to analyse your data in a matter of seconds. All you need to do is to input your data from the data summary form (see Table 8.2) into the software spreadsheet and then select the appropriate statistics to carry out the calculations.

Step 5: Decide whether result is significant (depends on significant level set)

Whichever type of statistical test you use, you will end up with a figure which is calculated in step 4 above. You then need to look in the statistical tables appropriate for that particular test to find out where your calculated figure falls. Each statistical table contains critical values against which the calculated figure can be compared. If your calculated figure is less than the critical value, then the results of your research are not significant and therefore accept your null hypothesis. If your calculated figure is equal or more than the critical value in the statistical table, then the results of your research are significant and therefore reject the null hypothesis.

Appendix 4 gives four different statistical tables which correspond to the four statistical tests that will be described later. In certain statistical tables such as the t distribution and the chi-square you will find the expression 'df'. This means 'degree of freedom'. Degree of freedom is a dimension that is used to find the critical value in the statistical tables. Sometimes it is determined by the number of subjects (respondents), sometimes by the number of subjects minus 1, and sometimes by the number of categories. However, you need not worry too much about the notion of degree of freedom or how to use the statistical tables. You will be guided on exactly what to do whenever degrees of freedom come up in connection with a statistical test.

There is one further point about the statistical inferential method and that is whether the research hypothesis is 'one tailed' or 'two tailed'. The point is that for a hypothesis which predicts a difference in only one direction (one tailed), there is a specified percentage probability that the difference might occur by chance but, if your research hypothesis makes a prediction that a difference might occur in either direction (two tailed), then there is double the probability that such differences might occur by chance. In the two-tailed condition there is the probability that a difference might occur in one direction plus the probability that a difference might occur in the other direction. However, you do not need to worry too much about the one-tailed and two-tailed notions, you can always use the probability level of the one-tailed direction.

Note to students: The statistical tests discussed below will provide you with a probability that allows you to judge whether your results are significant or are due to chance. Later, you will find out that all the worked examples will end up by stating the probability figure of the statistical test. For instance, you will come across an expression $P < 0.05$. This means that the probability of a result being due to chance is less than 5 per cent or 5 in 100. Another, $P < 0.01$, means that the probability of a result being due to chance is less than 1 per cent or 1 in 100, and so on. Therefore, the less the probability figure the more confident you can be in concluding that there is a significant difference applicable to your data.

The conventional level by which you can reject the null hypothesis and conclude that the results are significant is, $P < 0.05$. Otherwise, the results of your test are not significant and you have to accept the null hypothesis of no relationship or association between the research variables or issues under investigation.

Parametric test (the t-test)

The t-test is a parametric test which is used to compare the difference between the mean scores of two samples.

Example

Table 8.6 gives results of the unit cost of two samples of projects completed over the same period of time and they were of a similar type. All projects were office

Table 8.6   *Cost per square metre obtained from design and build projects and traditional projects*

Sample 1 (design and build projects) Unit cost (£) (X1)	Sample 2 (traditional projects) Unit cost (£) (X2)
564.00	505.00
521.00	557.00
445.00	465.00
560.00	458.00
480.00	530.00
540.00	480.00
585.00	665.00
426.00	525.00
530.00	605.00
475.00	495.00
$\bar{X}1 = £512.60$	$\bar{X}2 = 528.50$

blocks, reinforced concrete structure and of medium-sized projects, costing between £5 million and £10 million. However, one sample was taken from projects delivered using the design and build method of contracting and the other sample was taken from projects delivered using the traditional method of contracting. The aim of the research was to determine whether unit costs of Sample 1 (design and build projects) differ significantly from those of Sample 2 (traditional projects). Following the directions given in the above section, the following steps were taken:

1   Formulate the research hypothesis: 'The unit cost of projects delivered under the traditional form of contract (X1) is higher than similar projects delivered under the design and build contracts (X2).' Hi states that X1 > X2.
2   By implication, the null hypothesis is: 'There is no significant difference in the unit cost between the two samples.' Ho states that X1 = X2.
3   The t-test is the most suitable test for this type of problem. The rationale for using this test is that:

   (a)   The data is of an integer type (see definition in Chapter 7).
   (b)   The problem is to compare the mean of two samples.
   (c)   The samples are normally distributed. Note that, for the level of students reading this book, normality is not a strict condition.

4   Carry out the calculation (if not using the computer). In order for the mean unit cost of the two samples to be significant, the calculated value of 't' has to be equal or larger than the critical values of 't' given in Table A in Appendix 4. There are a number of formulas by which t can be calculated. This depends on whether the sample is small or large and whether it covers related or unrelated subjects. To simplify matters for students reading this book, 't' will be calculated by applying the following formula:

$$t = \frac{\bar{X}1 - \bar{X}2}{\sqrt{\dfrac{SD1}{n1} + \dfrac{SD2}{n2}}}$$

where

$\bar{X}1$ = mean for sample 1 $\bar{X}1 = 512.6$
$\bar{X}2$ = mean for sample 2 $\bar{X}2 = 528.5$
SD1 = standard deviation for sample 1   SD1 = 50.8
SD2 = standard deviation for sample 2   SD2 = 61.9
n1 = number of subjects in sample 1   n1 = 10
n2 = number of subjects in sample 2   n2 = 10

The standard deviation can be calculated using the formula given in the above section. However, most calculators can work out the SD in a matter of seconds so you can use it here. Therefore, in this example

$$t = \frac{512.6 - 528.5}{\sqrt{50.8/10 + 61.9/10}} = 2.05$$

5   Look up the significance of t in Table A of Appendix 4, taking into account the degrees of freedom, to see whether your result is significant.

For the t-test,

$$df = (n1 - 1) + (n2 - 1)$$
$$= (10 - 1) + (10 - 1) = 18$$

Look down the left-hand column of Table A to find the df (in this example 18). Then look along the row to see whether the calculated value of 't' is larger or equal to the critical values in the statistical table. In this example t = 2.05 which is just smaller than the critical value of 2.10.

From the above findings, one can say that the null hypothesis (of no difference) has to be accepted and conclude that there is no significant difference in the unit cost between projects delivered using the design and build contracts and those using the traditional form of contracts.

Note to students: If the calculated value falls between two critical values in the statistical table, say t = 2.4, you should use the expression .05 < P < .02 for a one-tailed hypothesis.

Non-parametric test (the chi-square test)

The chi-square (symbolised as χ^2) is a test of association between two sets of data and can be used only when the data is nominal or ordinal. It predicts how many subjects in each group will fall into certain categories. Therefore, you have to test quite a number of subjects to make sure that a sufficient number of subjects turn out to be allocated to each category. Most statisticians consider 20 subjects and over to be the minimum number of subjects required in order to apply the chi-square test.

Example

This example is related to Proposal 1 in Appendix 1, i.e. to find out whether companies who performed well financially also had a business philosophy different from those who performed badly financially. Table 8.7 shows the data that was gathered from 37 companies and were categorised as high financial performance (given numerate 2) and low financial performance (given numerate 1). The measurements of high and low were based on the annual profit of the companies. On the other hand, the business philosophy of the 37 companies was categorised as having a marketing-oriented philosophy (given numerate 2) and production-oriented philosophy (given a numerate 1). The question at issue here is whether business philosophy and financial performance 'go' together or whether there is no association between the two variables. Following the direction given above, the following steps were taken.

1 Formulate the research hypothesis: 'The type of business philosophy can have an effect on the financial performance of the private house-builders.'
2 By implication the null hypothesis is: 'The financial performance of private house-builders who have a marketing-oriented philosophy does not differ from those who have a production-oriented philosophy.'

Table 8.7 *Business philosophy and company performance*

Company	Business philosophy	Financial performance	Company	Business philosophy	Financial performance
1	1	2	20	1	2
2	2	2	21	2	2
3	1	2	22	2	1
4	2	2	23	2	2
5	1	2	24	1	1
6	2	2	25	1	1
7	1	1	26	1	1
8	2	1	27	2	1
9	2	2	28	2	2
10	2	2	29	1	1
11	2	2	30	1	1
12	2	2	31	1	1
13	1	2	32	1	1
14	2	2	33	2	2
15	2	2	34	1	1
16	2	2	35	2	2
17	2	1	36	2	1
18	2	1	37	2	2
19	2	2			

3   The chi-square test is appropriate to this kind of problem. The rationale for using this test is that:

(a)   The data is nominal (see definition in Chapter 7).
(b)   The research subjects (private house-builders) can be allocated into categories. In this example there are two categories.
(c)   The sample is quite large (37 subjects).

4   Carry out the calculation (if not using a computer). In order for the financial performance of the two categories to differ significantly, the calculated value of χ^2 has to be equal or larger than the critical values of χ^2 given in Table B of Appendix 4. To calculate χ^2 manually you need to take the following steps:

(a)   From the data summary sheet in Table 8.6, allocate the replies into one of the two categories, high financial performance or low financial performance. The results are shown in the form of a 2 × 2 table known as a 'contingency' table. The cells representing each of the categories are numbered 1 to 4 in Table 8.8. Remember that the figures in the table represent the number of subjects that fall into each category; they are not scores or percentages. For example, in the first top right cell, the number 6 means that in the sample, there are six companies who performed low financially and also have a marketing-oriented philosophy (check the rest by yourself).

Note to students: The rationale of the chi-square test is that it compares the 'observed' frequencies (O) in each of the squares (cells) of the contingency table with the 'expected' frequencies (E) for each cell if the differences are due to chance, as stated in the null hypothesis. In other words, the test compares the actual numbers of subjects (private house-builders) who fall into each cell as against the numbers of subjects we would expect to fall into each cell if there were in fact no differences between the financial performance of the two types of subjects (i.e. the marketing-oriented companies and the production-oriented companies). The observed frequencies are already known from the investigation (as shown in Table 8.8). Therefore, the expected frequencies need to be calculated. If the observed frequencies are close to the expected frequency, then the results are not significant. But, if the differences between the observed frequency and the expected frequencies are large, then the research hypothesis is supported. Therefore, the larger the χ^2, the more significant are the results, so the computed value of χ^2 should be equal or larger than the critical values in Table B of Appendix 4.

(b)   To calculate the expected frequency (E) for each cell, follow the procedure below:

   (i)   Add the total observed frequencies for the rows and columns. In this example, they are equal to 23, 14, 22, 15. Here, the grand total is 37 (23 + 14 for rows) and (22 + 15 for columns) – see Table 8.8.

   (ii)   Multiply the two relevant marginal totals for each cell and dividing by the total number of subjects N.
          Therefore,

          For Cell 1: the expected frequency (E) = 22 × 23/37 = 13.67
          Cell 2: E = 15 × 23/37 = 9.32
          Cell 3: E = 22 × 14/37 = 8.32
          Cell 4: E = 15 × 14/37 = 5.67

Table 8.8   *Contingency table between business philosophy and financial performance*

	High financial performance (2)	Low financial performance (1)	Total
Marketing oriented (2)	**1** 17 E = 13.67	**2** 6 E = 9.32	23
Production oriented (1)	**3** 5 E = 8.32	**4** 9 E = 5.67	14
Total	22	15	37

Chi-square = 5.26

(iii) Find the value of chi-square from the formula $\chi^2 = \Sigma(O - E)^2/E$, where,

O = observed frequencies for each cell (as shown in Table 8.8)
E = expected frequencies for each cell (as calculated above)
Σ = add up the results of $(O - E)^2/E$ calculated for each cell.

Therefore,

$$\chi^2 = \frac{(17 - 13.67)^2}{13.67} + \frac{(6 - 9.32)^2}{9.32} + \frac{(5 - 8.32)^2}{8.32} + \frac{(9 - 5.67)^2}{5.67}$$

$$= 0.81 + 1.18 + 1.32 + 1.95 = 5.26$$

5. Look up the significance of χ^2 in Table B of Appendix 4, taking into account degrees of freedom, to see whether your result is significant.

> For the χ^2, the degree of freedom (df) $= (r-1)(c-1)$, where,
> r = number or rows in contingence Table 8.8 (r = 2)
> c = number of columns in contingency Table 8.8 (c = 2)
> Therefore df $= (2-1)(2-1) = 1$

> Table B gives critical values against which the calculated value of χ^2 can be compared. The level of significance depends on the degrees of freedom (df). In this example, df $= 1$ so we should look along that row in Table B. Since our calculated value of χ^2 is 5.26 and so is larger than the critical value of 3.84 for P $< .05$, the results of this research are significant.

From the above finding one can say that the null hypothesis (of no difference) has to be rejected and conclude that there is a significant difference in the financial performance between house-builders who had a marketing-oriented philosophy and those who had a production-oriented philosophy.

Notes to students:

- In this example, the calculated value of χ^2 falls between two critical values in the statistical table (between 0.05 and 0.02). In this case we use the expression $0.02 < P < 0.05$.
- Remember, it is absolutely essential to make comments on and interpretations of your findings after the analysis. What does this finding tell you? What does it mean? Does it correspond closely with previous studies or not. Chapter 9 provides you with a guidance on how to comment and interpret your results.

Spearman rank correlation coefficient

The Spearman (rho) correlation is a non-parametric test for measuring the difference in ranking between two groups of respondent's scoring a number of issues, attributes or factors. The test can be applied when:

1 The question is requesting a ranking value like that on pp. 81–82.
2 The question is requesting a scoring value and the scores are then converted into rankings like that on p. 80.
3 The question is requesting rating values and the ratings are converted into rankings like that on pp. 76, 78.

Example 1

Consider the question on p. 81. The researcher sent the questionnaire to site operatives of two different industries, i.e. to 50 workers in Indonesia and to 50 workers in the UK. The student wanted to find out whether there is a difference in attitude between the two groups towards the factors that can affect job satisfaction. Forty-two replies were received from UK workers and 44 replies from Indonesian workers. Table 8.9 shows the data obtained. Column

Table 8.9   *Average ranking of UK workers and Indonesian workers to factors influencing job satisfaction*

Job satisfaction factors	UK ranks (1)	Indonesia ranks (2)	di (3)	di^2 (4)
1 A well-paid job (having a good income and being able to afford the good things in life)	3	1	2	4
2 A prosperous job (having a good opportunity to develop myself)	1	5	4	16
3 An important job (a job full of achievement that brings me respect and recognition)	4	7	3	9
4 A secure job (making certain that I have steady work and steady wages)	2	2	0	0
5 Relationship with fellow workers (getting along well with the people I work with)	8	6	2	4
6 Relationship with superiors (getting along well with my supervisor)	6	4	2	4
7 Job interest (good chance to do work of my interest)	5	3	2	4
8 Working condition (working within a good physical job condition)	3	8	5	25
Total di^2				66

(1) represents the average rank for UK workers and column (2) represents the average rank for the Indonesian workers. Columns (3) and (4) represent the difference in ranks and their squares respectively. The following steps were taken to apply the appropriate test:

1   Formulate the research hypothesis: 'The factors that satisfy workers with their jobs differ between Indonesian workers and UK workers.'
2   By implication, the null hypothesis is: 'There is no significant difference in the ranking of job satisfaction factors between the two industries.'
3   The Spearman rank correlation is an appropriate test to this type of problem. The rationale for using this test is that:

  (a)   The attitude is measured on an ordinal scale (see definition in Chapter 7).
  (b)   The problem is to measure the amount and significance of a correlation between people's rank on a number of issues.

4   Carry out the calculation (if not using a computer). Here, the prediction is that, if there is a high positive correlation in opinions between the two groups, then the low rank of one item by the UK operatives should correspond to low rank by operatives in Indonesia and vice versa. However, if there is no correlation (as stated by the null hypothesis) the average ranks will be mixed up since one group could have a high ranking on the eight-point scale while the other have a low ranking for the same item. To calculate (rho) you need to apply the following simple formula:

$$\text{rho} = 1 - \frac{6 \, \Sigma \, d_i^2}{N(N^2 - 1)}$$

where, d_i = the difference in ranking between each pair of factors
  N = number of factors

Therefore, in this example,

$$\text{rho} = 1 - \frac{6(66)}{8(64 - 1)}$$

$$= 1 - \frac{396}{504} = 0.21$$

5   Look up the significance of (rho) in Table C of Appendix 4, the calculated rho value of 0.21 against the value N (number of factors); in our example N = 8, is less than the critical value of 0.643 for P < 0.05 in the one-tailed test. Therefore, you can conclude that there is no correlation between the attitude of both groups and the null hypothesis should be accepted. This means that most of the factors that are considered very important to satisfy the UK worker are less important to workers in Indonesia and vice versa.

Note to students: The value of N will be taken as 'number of subjects' if the question or the investigation is of a different format as in the case of an experimental study. In construction, an example could be when 30 people are asked to rate two different attributes where N will be calculated as the number of subjects that took part in the research which, in this case, is 30 (see example in section 9.3 of Green and D'Oliveirs' 1993 book).

Example 2

In this example I will discuss a situation when the question is requesting a rating value and the ratings are converted into ranks. Here, the student asked two groups of people, namely, head office personnel and site managers, to respond to a number of questions by ticking their opinion on a three-point scale. Thirty-six fully completed questionnaires were returned, 19 from contract managers working at head office and 17 from experienced site managers. After receiving the data and recording the information in the data summary sheet, the student assigned scores to the ratings. Highly important = 3 points, moderately important = 2, and less important = 1. Table 8.10 shows results of the survey. Here, the student applied a simple arithmetic to convert the average ratings into percentages of actual scores. Hence, Table 8.10 was constructed calculating the maximum possible scores, actual scores and percentage of actual scores for head office respondents and site managers. The maximum possible score for each factor relating to head office respondents is $3 \times 19 = 57$ and for site managers is $3 \times 17 = 51$. For example, ten of the head office personnel thought that 'project planning' can highly affect productivity, eight thought moderately and one did not think it has any effect at all.

Table 8.10 *Ranking of head office personnel and site managers to factors influencing site productivity*

Productivity factors	Head office actual scores (%)	Rank (A)	Site managers actual scores (%)	Rank (B)	di (A − B)	di²
1 Delegation of responsibilities	58	24	70	14.5	9.5	90.2
2 Integration of information	77	5	83	3	2.0	4.0
3 Project planning	82	1	85	1	0	0
4 Scheduling of project activities	75	9	75	14.5	5.5	30.2
5 Level of authority	53	27.5	52	29	1.5	2.2
6 Supervision of subordinate	75	9	73	11	2.0	4.0
7 Communication	60	22	78	8.5	13.5	182.0
8 SM role in contract meetings	75	9	73	11	2.0	4.0
9 Attitude of site personnel	75	9	64	20	11.0	121.0
10 The decision-making process	75	9	64	20	11.0	121.0
11 Procurement of materials	77	5	83	3	2.0	4.0
12 Site programme	77	5	78	8.5	3.5	12.2
13 Accuracy of tech. information	67	14.5	81	6	8.5	72.2
14 Fitness of tools for tasks	67	14.5	70	14.5	0	0
15 Knowledge of technology	67	14.5	81	6	8.5	72.2

16	Interference on workmanship	67	14.5	73	11	3.5	12.2
17	Site layout	63	20.5	62	24.5	4.0	16.0
18	Promotion of employee	53	27.5	64	20	7.5	56.2
19	Resentment of company policy	65	18	64	20	2.0	4.0
20	Incentive for good performance	55	26	62	24.5	1.5	2.5
21	Opportunities to exercise skill	39	29	68	17	12.0	144.0
22	Response to employee grievances	58	24	62	24	0	0
23	Uncertainty about career	72	12	70	14.5	2.5	6.25
24	Work environment	81	2	83	3	1.0	1.0
25	Experience of employees	79	3	81	6	3.0	9.0
26	Contract administration skill	65	18	64	20	2.0	4.0
27	Knowledge of techniques	63	20.5	62	24.5	4.0	16.0
28	Training on new technology	58	24	59	27.5	3.5	12.2
29	Availability of skilled personnel	65	18	59	27.5	9.5	90.2

Total di^2 1092.7

Note: $\text{rho} = 1 - \dfrac{6\sum di^2}{N(N^2 - 1)} = 1 - \dfrac{6 \times 1092.7}{29(841 - 1)} = 0.73$, hence, $P < 0.005$

The actual scores of head office personnel to 'project planning' has then been calculated as:

$$(10 \times 3) + (8 \times 2) + (1 \times 1) = 47$$

Therefore, the percentage of maximum scores was

$$47/57 \times 100 = 82 \text{ per cent}$$

Table 8.10 shows the complete results. The ranking of the 29 factors between the two groups was tested using Spearman's rho, with the correlation found to be significant at $P < 0.005$ (see calculation for rho $= 0.73$ in the note to Table 8.10). This indicates that high ranking to one productivity factor by one sample, in most cases, corresponds to high ranking by the other sample and vice versa. The research hypothesis, therefore, can be rejected, and the null hypothesis can be accepted, concluding that the difference in opinion between head office personnel and site managers towards factors influencing productivity is not significant.

Pearson product moment correlation coefficient

The correlation coefficient (r) is a parametric test which is used when you have two sets of scores and you want to calculate whether there is a strong relationship between them. Whereas the Spearman test can be used with any data, the Pearson requires the research data to be measured on an 'interval' scale and to meet the other assumptions for parametric tests. With the Pearson correlation, the relationship can be either positive or negative and the strength of the relationship is measured on a scale that varies from $+1$ through 0 to -1. When one variable increases as the other increases the correlation is positive; when one decreases as the other increases it is negative. The variable can be either factual or perceptional. For example, we would say that the speed of construction is correlated with the unit cost of the building. Since one increases with the other the correlation is called positive. In contrast there is a negative correlation between complexity of the building and speed of construction. In other words, the more complex the project, the slower the rate of construction.

After you collect the data, it is best to draw a scatter diagram which is a graph with axes X and Y corresponding to the two variables being measured. The two variables can represent two dependent factors, two independent factors, or one dependent factor and one independent factor. The scatter diagram should show at a glance whether a relationship is likely to exist or not. Figure 8.5 shows a number of examples.

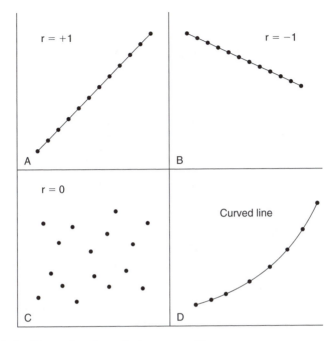

Figure 8.5 *Examples of correlation scatter diagrams*

Example

Let us assume that Table 8.11 is a set of data collected by the researcher and he/she wants to assess whether there appears to be a correlation between the two measurements labelled X and Y. For instance, X could be the mathematics marks of the students out of 100 per cent, Y could be the physics marks out of 100 per cent. The question at issue is to find out whether maths performance and being good at physics 'go' together or whether there is no connection between the two kinds of ability.

In order to calculate the significance of the correlation, the following steps were taken:

1 Formulate the research hypothesis: 'Students who are good at maths are also good at physics, students who are average at maths are also average at physics and students who are poor at maths are also poor at physics.'
2 By implication, the null hypothesis is: 'There is no significant relationship between the maths results and physics results.'

Table 8.11   *Scores on variable X and Y for 12 subjects*

Number of subject	Variable X	Variable Y	(X) × (Y)	X^2	Y^2
1	65	68	4420	4225	4624
2	63	66	4158	3969	4356
3	67	68	4556	4489	4624
4	64	55	3520	4096	3025
5	68	69	4692	4624	4761
6	62	66	4092	3844	4356
7	70	68	4760	4900	4624
8	66	65	4290	4356	4225
9	68	71	4828	4624	5041
10	67	67	4489	4489	4489
11	69	68	4692	4761	4624
12	71	70	4970	5041	4900
Total	800	801	53467	53418	53649

3   The Pearson product moment correlation is an appropriate test to apply. The rationale for using this test is that:
   (a) The data is of an interval type (see definition in Chapter 7).
   (b) The problem is to establish a relationship between two variables.
   (c) The sample is normally distributed. Note that, for the level of students reading this book, normality is not a strict condition.
4   Carry out the calculation of the correlation coefficient (if not using the computer). In order for the scores for the two variables to be highly correlated, the calculated value of 'r' has to be equal or larger than the critical values of 'r' given in Table D in Appendix 4. The correlation coefficient (r) can be calculated from the following formula:

$$r = \frac{N \Sigma XY - \Sigma X \Sigma Y}{\sqrt{[N \Sigma X^2 - (\Sigma X)^2][N \Sigma Y^2 - (\Sigma Y)^2]}}$$

$$r = 0.59$$

5   Looking up significance of (r) in Table D of Appendix 4, the correlation coefficient of 0.59 indicates a positive correlation between variable X (maths marks) and variable Y (physics). Looking up for significance in

Table D of Appendix 4, for df = (10 − 2), the calculated value of 0.59 exceeds the critical value 0.5494 for P < 0.05 (one tailed) and therefore the researcher can conclude that there is a significant relationship between the two variables. This means that there is less than 5 per cent probability that it was due to a chance distribution of scores. But in interpreting correlation it is important to remember that 'correlation is not causation'. There may or may not be a causation connection between the two correlated variables. However, for the level of students reading this book, it will be enough to comment on what was stated in your research hypothesis. In this example, you can say that the research hypothesis is valid and, therefore, reject the null hypothesis, concluding that: 'Students who perform well in maths also perform well in physics.' If your dissertation contains a number of findings such as these, then it can be said that your dissertation has a substance.

Summary

This chapter explained the methods by which you can analyse the results of your investigation. It first explained the descriptive method and then the inferential method with worked examples.

The descriptive method of analysis includes the frequency distribution, measurement of central tendency and measurement of dispersion. With the descriptive method you can apply a straightforward calculation to show how the respondents are distributed on all the items of the investigation. For example, 20 per cent of the house-builders believed that their firm's business philosophy is to 'first build the estate and then work out how best to sell the product'. Whereas, 80 per cent believed the opposite. Such breakdown of percentages or the numbers of observations that fall into each of several categories are termed 'frequency distribution'. The frequency distribution can be illustrated in a table, bar chart, graph or pie chart, all of which are demonstrated in this chapter.

Measures of central tendency are applied when you have a group of data and you wish to find the most typical value for the group or the score which all other scores are evenly clustered around. These statistics are known as the mean, mode and median.

Measurement of dispersion is a type of analysis that can show you the degree by which numerical data tend to spread about an average value and it is called variation, dispersion or standard deviation.

The inferential method of analysis concentrates on the nature of the relationship between two variables and on the construction of measures of relationship. Four types of tests were explained with examples in this chapter, these are t-test, chi-square test, Spearman (rho) ranking correlation and Pearson product moment correlation (r).

The t-test is a parametric test and it is appropriate for use when you want to compare the difference between the mean scores of two groups of data. The data has to be of interval type and normally distributed, and can be used, for example, to find out if there is a significant difference in the unit cost between two types of contracts.

The chi-square test is a test of association between two sets of data and can be used when the data is nominal or ordinal. It predicts how many subjects in each group fall into certain categories, for example, how many companies who are marketing oriented also perform high financially, and how many product-oriented companies also perform poorly financially.

The Spearman's ranking correlation (rho) is a non-parametric test for measuring the difference in ranking between two groups of respondent's, scoring a number of issues, attitudes or factors. It can be used, for example, to find out whether two groups of people, say, managers and operatives, are ranking their priorities against a list of job satisfaction factors. The question at issue here is whether there is a significant difference in their ranking.

The Pearson product moment correlation (r) is a test used when you have two sets of data and you want to calculate whether there is a strong relationship between them. The relationship can be either positive or negative and the strength of it is measured on a scale that varies from $+1$ through 0 to -1. For example, it can be used to find out whether speed of construction is related to size of project.

References and additional reading

Bell, J. (1996) *Doing Your Research Project: A Guide for First Time Researchers in Education and Social Science.* Open University Press.

Coolican, H. (1993) *Research Methods and Statistics in Psychology.* Hodder and Stoughton.

Green, J. and D'Oliveirs, M. (1993) *Learning to Use Statistical Tests in Psychology: A Student's Guide.* Open University Press.

Kennedy, J. and Neville, A. (1976) *Basic Statistical Methods for Engineers and Scientist.* 2nd edn. Harper and Row.

Nachmias, C. and Nachmias, D. (1996) *Research Methods in the Social Sciences*, 5th edn. Arnold.

Rassam, F. (1989) Marketing Profile and Financial Performance of Private House Building Companies. MSc dissertation, Bath University.

Siegel, M. (1995) *Theory and Problems of Statistics.* Schaum's Outline Series, McGraw-Hill.

Swinscow, T. (1987) *Statistics at Square One.* British Medical Association.

Zikmund, W. (1997) *Exploring Marketing Research*, sixth edn. Dryden Press.

9 Structuring and writing the dissertation

This chapter will explain the general format for writing, organising and presenting your dissertation. These formats will provide you with a guide to structure your dissertation which should make it easier for the reader to understand what your dissertation is all about. Specific requirements for presenting the dissertation may vary from one discipline to another and from one university to another. For example, the number of copies required, format of the front cover, size of paper, type-setting and the like. However, the general formats of presentation are much the same in most disciplines, and so the main task is to make sure that you write and organise your dissertation contents correctly.

This chapter will give you guidance on how to structure and write your dissertation. It gives examples of writing an abstract, an introduction, a piece of literature review, present the research design, present the chapter of findings and conclusions. The examples given in this chapter should be treated as guidance which can serve as a model only. The contents of Chapter 9 are shown in Figure 9.1.

Typical dissertation structure

The structure of your dissertation may take the following order:

1 A page showing the title, your name, department and a copyright statement.
2 Summary of figures (if any).
3 Summary of tables (if any).
4 Acknowledgements.
5 Abstract.
6 Introduction to the research. One chapter.
7 Literature review. Usually two or three chapters reviewing previous work (theory/ies and/or previous evidence).

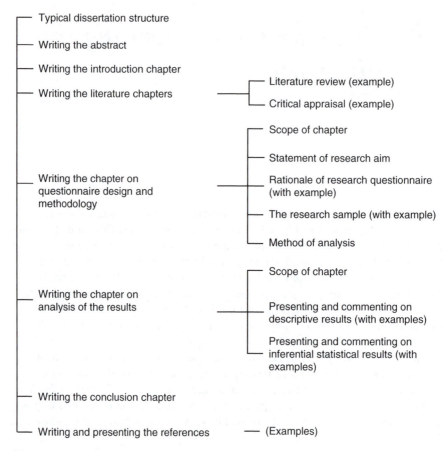

Figure 9.1   *Contents of Chapter 9*

8 Questionnaire design and method of analysis. One chapter. If the dissertation is purely based on analysing archival data, then you may name the chapter 'Procedure of data gathering'.

9 Analysis of the results. One chapter. If the dissertation is based on appraising previous work, then you may name it 'Analysis of secondary data' or 'Critical appraisal of previous work'.

10 Summary and conclusions.

11 Recommendations for further studies and practice.

12 References.

13 Appendices.

Writing the abstract

The abstract is a brief summary of the research, usually about 200 words in length. It should give information on the research problem that was studied, the method used to study the problem (including information on the type of sample), the results, and major conclusions. The abstract should provide enough information to the examiners about your research as well as information to other readers so that they may decide whether to read the entire dissertation. The abstract should also make the dissertation easier to comprehend when it is read. Although the abstract appears at the beginning of your dissertation, you will probably want to wait until the body of the dissertation is complete before you write the abstract.

You may need several attempts before you achieve a sufficient brief. The following example fulfils all the requirements of an abstract, and might serve as a model.

Abstract (example of writing up style)

This dissertation is about evaluating the perception of head office personnel and site managers into the factors that influence productivity on site. A critical review of the literature is structured under three chapters: (1) management; (2) motivation; and (3) experience and training. Twenty-nine factors were extracted from the above headings and were assessed by 19 head office personnel and 17 site managers. The survey indicated that both samples regard 'ineffective project planning' and 'constraints on a worker's performance' as the most critical factors influencing productivity. Another highly ranked factor by both samples is 'lack of integration of project information'. Given the large number of computer systems available, this must be disheartening and reflects similar findings of recent research in the use and application of IT systems.

Writing the introduction chapter

This chapter introduces the reader to the area being investigated. It basically includes most of your proposal contents. After reading the introduction, the reader should know why you decided to do the research and how you decided to go about it. In general the introduction chapter may be composed of four sections, these are:

1   The first section starts with a description of past research which is relevant to the problem. This section may be entitled 'Nature of the problem', 'Scope of the research', 'Rationale for the research', 'Background to the problem'

and the like. Here, an exhaustive review of past research is not necessary, rather you want to describe only the research and/or theoretical issues that are clearly related to your study. You should state explicitly how this previous work is logically connected to your problem. This tells the reader why your research was conducted.

2   The second section of the introduction tells the reader what your aim is, what objectives are to be achieved and what hypothesis (or key questions) are being examined. This part may be entitled 'Research goals'.

3   The third section should describe the method you used to achieve your aim and objectives. This part may be entitled 'Outline methodology of the research'.

4   The final part of the introduction tells the reader about the chapters of the dissertation. This part may be entitled 'Dissertation contents'.

Writing the literature chapters

The literature review is usually spread over two or three chapters. These chapters review what has been written on your topic (see Chapter 2). Haywood and Wragg (1982, p. 1) emphasised that a critical review should show that 'the writer has studied existing work in the field with insight'. The main point is that a review should provide the reader with a picture of the state of knowledge and of main questions in the subject area being investigated. When writing the literature review chapters make sure of the following:

1   Sifting the information. This means that you should only include materials which are directly related to your study.

2   Clarity in writing. Be precise and clear in presenting other people's work.

3   Coherence in writing. The writing of the literature and indeed the whole dissertation should be coherent. The material should be presented in an orderly, logical progression to facilitate understanding.

4   Ensure grammar, punctuation and length are correct.

5   Acknowledge the work of others. It is extremely important to clearly separate your own words and ideas from those obtained from other sources. In most research reports including a dissertation, almost certainly all the information provided in the literature chapters are drawn from previous work. Therefore almost every statement you make in the literature review should be referenced. If you use a passage drawn from an article or book, make sure that the passage is presented as a direct quotation. There is nothing wrong with quoting another author as long as you acknowledge your source (see Referencing on p. 153).

The following example might serve as a model to show sifting, clarity and acknowledgements. The literature review is extracted from a study by Naoum (1989, pp. 67–72) entitled 'Models of the building process'. You may not be familiar with the field of study, but the review shows what is expected from you. Naoum reviewed several models of the building process and then selected the ones which dealt with integration and the system theory and ignored others, i.e. he sifted the information which he gathered. He compared previous models and discussed in some detail a study by Sidwell which served as a pilot for his more extensive study of developing a 'building process model'.

Literature review (example of writing up style)

Since the 1960s various models have been developed to investigate the effectiveness of the building team or the operation of the building process. Some researchers developed their model to show the structure of the industry such as the Tavistock model, but others were interested in showing the processes and participants involved in project development and have a temporal aspect. In other words, they show sequences of events and not instants in time. However, although the Tavistock report produced a sequential type of model, it did open the building industry's eyes up to a system approach. According to Cleland and King (1968), the system approach illustrates the interaction and interdependence between the identified variables, suggesting that an action of one variable can cause reaction on the part of others.

Among the most relevant models for this research that followed the Tavistock report are those of Morris (1972), Walker (1982), Ireland (1983), Sidwell (1982), Nahaphiet (1985), Wearne (1984), Rowlinson (1988) and Newcombe (1988).

Morris (1972) took a system view of the building industry and studied the interfaces between the design and production. He used the Tavistock report and the work of Miller and Rice on organisational boundary definition (1967) and Lawrance and Lorsch's studies of differentiation and integration (1967) as a mainspring of his research model. Morris broke down the building process into three main subprocesses which may occur sequentially or concurrently i.e. outline design, detailed design and construction. The concurrence of the building process is an indicator of the degree of integration of the building process. This integration may be modified by the managerial actions which determine the make up of the building team and the parameters by which they are guided. This managerial action may in turn be modified by the environment, which constitutes all the factors which influence the client, the building team and the building process. *This provides one of the first pieces of work which approaches the problem in terms of the process as a whole and attempts to provide a rationale for actions.*

Walker (1980) also adopted a system viewpoint and defined a model which is client oriented and is common to all projects. He used the technique of linear responsibility analysis to investigate decision making and appropriate organisation structures for construction project management. The model is in terms of three stages: project conception, project inception and project realisation. *In recognising the non-sequential nature of the construction process, the decision points within the system adds tasks discontinuity to Miller's work (1956) on technology, territory and time.*

Ireland (1983) adopted Kast and Rosenweig's model (1973) of the organisation and indicated that he had reversed their proposition of management and structure being dependent systems. Ireland conducted his research on the basis that 'Technology used, structure chosen, the psychosocial aspects and the way the project is managed, will all have an effect on the achievement of goals and values subsystems'. Ireland maps these sub-systems to form a strategic control of the building process but omits discussion of who should exercise this control. *It also appears that the concept of socio-technical analysis is not fully supported from his research. This is so due to the cross-sectional approach to the research method evidence for impact of the social system and the adaptive controls would be very difficult to find.*

The various models of previous researchers were studied by Sidwell (1982) who criticised them however for they infer a sequential process and do not illustrate the interactive and cyclic nature of the building process. He, therefore, identified and studied the interrelationship between 19 variables and discussed them under six main headings. These are: 1) client characteristics, 2) project characteristics, 3) project procedure, 4) building team, 5) environment 6) project success. Sidwell applied the model over 32 case studies to examine, on the one hand, the relationship between the variables: client, project building team and project procedure, which together define the organisational form. And on the other hand, variables of the organisational form are examined with variables of project success in respect to time and cost.

Naoum went on and stated that, Sidwell's model (1982) did not include designer characteristics nor client and contractor criteria, which are modified by the foregoing, in order to gauge performance. These criteria have been assumed previously and deemed to be project characteristics. Having said that, the success measures that he used were both subjective and objective which helps counter criticism on criteria.

6   Appraisal of the literature chapter (approximately 500 words). *At the end* of each literature chapter you need to include an 'appraisal' section to discuss the implications of the information provided in that particular chapter on your investigation. Sometimes it is called 'critical appraisal', 'Appraisal', 'Discussion', 'Summary', 'Concluding remarks' and the like. In this section, you are allowed to discuss or criticise other's work using your own words. The following appraisal section is taken from a dissertation by Stuart Green (1996, pp. 34–36) on health and safety in construction. The following example might serve as a model.

Appraisal of the literature chapter (example of writing up style)

This chapter set out to establish whether the construction industry's safety record is poor when compared to other major UK industries. A further objective of the chapter was to analyse the accidents rates and try to offer an explanation as to why these accidents occur. The first part of the chapter looked at the history of construction site accidents and showed that employment was falling (20% on the 1981 figure) and so was the incidents of accidents. Self-employment has risen by 45% over the last ten years and this has also seen a decline in the accident rate. Self-employment continues to be a grey area with regards to health and safety, but overall the combined (employed and self-employed) fatalities fell by 16% to 4.7 per 100,000 workers.

The chapter then went on to look at how at the construction industry compared with other industries with regards to health and safety. The section described how the reporting of accidents will rise with the introduction of new reporting measures but the trend will continue downwards.

Accidents rates for all industries have fallen by 25% since 1993/94, due mainly to the increase in employment in the safer industries. The construction industry saw a 20% fall in fatal injuries from 1993/94, with energy and manufacturing showing a 51% and 29% reduction respectively. The difference in accident rates between construction and manufacturing may attribute to four points as follows:

1   Factories are controlled environments
2   There is little change in manufacturing procedures or equipment
3   The manufacturing labour force remains constant
4   Hazards in factories can be remedied easier and with little cost

Accidents on construction sites usually cause multiple fatalities and injuries, such as the Ashford Disaster in the summer of 1995 when a three story building collapsed during refurbishment works killing four site workers (Construction News: 25,4,96). Accidents in factories rarely involve more than one person.

The final section of this chapter looked at accident trend in the construction industry. The figures showed that fatal and major injuries were mainly cased by fall from a height. Accident rates continue to fall but fatalities for falls remain constant. The most hazardous occupations continue to be roofing and labouring, probably due to the large amount of untrained labour employed in these sectors.

In conclusion, contents of this chapter confirm that the construction industry is risky as far as health and safety is concerned. However, the trend for accidents is downwards and the implementation of new safety legislation such as CDM, and the increasing awareness of management, is likely to improve safety although there may be a financial cost associated with the improvement.

The following chapter explains the cost implications as a result of improving safety in the construction industry.

Writing the chapter on questionnaire design and methodology

This chapter needs to be inserted after the literature chapters. It may also be entitled 'Procedure of data gathering', 'Research design and methodology' or the like. In this chapter, you need to explain how the problem was investigated and describe the tool that you used to make the investigation, i.e. your questionnaire or the data collection format. You also need to describe the characteristics of your research sample and method of analysis. I suggest that this chapter may include the following sections.

Section 1: scope of chapter

This should be a brief explanation of what the reader will expect to read in this chapter.

Section 2: statement of research aim

This should be a repeat of your stated aim to remind the reader as to what you intend to investigate.

Section 3: rationale of the research questionnaire

In this section you need to rationalise your questionnaire by stating why have you asked the questions you did. Remember, your questionnaire consists of, say, three to five section headings and each heading contains a number of questions which are related to your research objectives. Your task, therefore, is to refer to the questionnaire which is usually inserted in the Appendix and discuss each section individually. For instance, what did you intend to obtain from the respondent? Which objective would you achieve by asking that group of questions? Has there been a similar type of question asked by fellow researchers? If so, refer briefly to their questionnaire or to their work. For example, in the study related to Appendix 1 on total quality management, one of the sections was entitled 'Key components of service quality'. The student wrote the following rationale which might serve as a model (Hania Cooper, 1995, pp. 89–91).

Questionnaire rationale: key components of service quality (example of writing up a rationale)

This section of the questionnaire lists and defines the functions of fifteen 'key components' of quality. The fifteen components comprise an adaptation of those proposed by Garvin (1987), Haywood-Farmer (1987) and Wilson (1994), amended by the author to suit the context of a multi-disciplinary professional service firm in the construction industry.

Previous chapters have suggested that a standard accredited Quality Assurance (QA) system such as ISO 9001 may not sufficiently address all the pertinent aspects of quality within a professional service and it has been proposed that a market-focused approach to quality would best satisfy client requirements (Barrett 1992). It is submitted that such an approach might commence by isolating the 'key components' of service quality, such as 'competence of personnel' and 'reliability of service'. It would then be possible to establish the relative importance of such key components to the client and then use performance data to determine how far service performance matches clients expressed needs (Wilson 1994).

The company surveyed did not extend QA to marketing functions as such and, in view of the understandable importance to the firm of confidentiality in this area, detailed marketing data was excluded from the questionnaire (note that, in this particular dissertation, the student used her own firm to investigate the issues of her research). Instead, it was felt useful to use the concept of 'key components' of quality to address the respondent's own perception. Following an informal pilot survey of three construction professionals, it became clear that respondents were likely to find it too difficult and time-consuming to rate or grade each of the fifteen 'key components'. Therefore, the question was changed slightly by asking the respondents to indicate the five most important and five least important 'key components'. List of the fifteen key components and their definitions are given below:

1   Performance – defined as the provision of the basic service such as project management or quality surveying.
2   Feature – defined as supplementing basic service such as building surveying supplementing project management.
3   Reliability – defined as consistency of performance, dependability and low risk of errors.
4   Responsiveness – defined as willingness to personnel to provide service such as giving prompt service.
5   Competence – defined as profession of required skills.
6   Conformance – defined as the extent to which service meets any established standard such as Housing Corporation Scheme Development Standard.
7   Access – defined as the approachability and ease of contact.
8   Currency – defined as the ability to keep relevance and up-to-date such as with new legislation, methods and products.

9 Courtesy – defined as politeness, respect, consideration and friendliness of contact personnel.
10 Communication – defined as keeping informed in language they can understand such as explaining reasons for delays or trade-offs between service and cost.
11 Perceived quality – defined as trustworthiness, believability such as reputation of the firm or personal characteristics of contact personnel.
12 Security – defined as freedom from danger, risk or doubt such as financial security and confidentiality.
13 Understanding – defined as knowing the customers; making the effort to understand the client's needs such as specific requirements and individual attention.
14 Tangibility – defined as the physical evidence of the service such as physical facilities and equipment.
15 Costing – defined as clear pricing or accurate estimating such as known financial commitment, no hidden extras.

Note to students:

1 If your research involves a theoretical framework, you need to give details of the variables or factors that you have studied. If you have designed a model, you need to give details of its components and state your argument as to how these components are interrelated. In other words, you should define the variables one by one and describe the method which you used to measure the variables.
2 If your investigation is based purely on archival data and not a questionnaire, then you need to discuss the format that you used to gather the information. Remember, the archival information can be either statistical or descriptive.

Section 4: the research sample

In this section of the chapter you need to discuss the following:

1 The list that you drew your sample from.
2 The method that you used to select your sample and why.
3 If you used a survey questionnaire, state how many were sent out and how many were completed, i.e. response rate.
4 Describe the characteristics of your response sample. This can be discussed from the information gathered by the 'general' section of your questionnaire. Table 9.1 shows an example. This part of the questionnaire was designed to establish the range of respondents in terms of annual turnover and to gain an insight into the role marketing plays within their business strategy.

Table 9.1   *Research sample: number and rate of response by category (example of writing up style)*

Category	Questionnaires issued	Responses	% Response
Directors	12 (14)	7 (14.5)	58
Associates	12 (14)	8 (17.0)	67
Constr. Profession.	55 (63)	26 (54.0)	47
Admin.	8 (9)	7 (14.5)	88
Total	87	48	55

Note: Figures in brackets indicate the corresponding percentage relative to total questionnaire issued (87 = 100%), or relative to total response (48 = 100%), as applicable to that column.

Section 5: method of analysis

This section of the research design chapter should explain how the data was analysed. If you selected the interview technique, have you used the trend approach in analysing the information or percentages? If you selected the postal survey, have you used the descriptive method of analysis or the statistical test/s?

Writing the chapter on analysis of the results

This chapter is the core part of your dissertation and will consist of the research findings expressed by texts, figures, tables, charts, graphs and the like. It is absolutely important to present your results clearly. For this matter I suggest the following structure for this chapter.

Section 1: scope of chapter

This section should be a brief description of what is included in the chapter.

Note to students: Chapter 8 in this book described the methods for analysing the results. If you decided to apply the descriptive method follow Section 2 part 1 below, if you decided to apply the statistical test, then follow Section 2 part 2 below and if you decided to apply both methods, then follow parts 1 and 2 below.

Section 2: analysis of the results, part 1 (the descriptive method)

This part of the dissertation will consist of questions, answers, tables, figures or charts and comments. When writing the results section, it is best to refer back to your questionnaire. The order in which your results are presented should correspond to the order of your questionnaire. In general, the analysis of the results for the descriptive method can be presented following the format below:

1   *The question.* Here, you state the question as was asked in the questionnaire.
2   *The results/finding.* Here, you present the analysis of the results in the form of table, bar chart, pie chart or a graph.
3   *Comments/interpretation.* After analysing the question and presenting the results in a table, chart or other form, you need to go beyond this step and interpret the results of your study. For instance, what do they mean to you? Are your results similar or dissimilar to the results of previous studies? If your results were different, why were they different? Did you use different subjects or use different techniques?

Below are three examples of a typical presentation of a descriptive type of results. The first example is when your investigation is based purely on a postal survey, the second when you conduct a postal survey supplemented by interviews and the third example is based purely on interviews.

Presenting a postal survey (example of writing up style)

Question:
This question is related to the adequacy of the 'current' Quality Assurance Procedures to satisfy clients' requirements and expectations. Here, the respondents were given a number of factors which can contribute to the quality of service to satisfy clients' requirements. They were asked to mark each factor with a figure '3', '2' or '1'. Mark '1' means that the 'current' QA meets the client satisfaction to this particular factor. Mark '2' means that it moderately meets client satisfaction and mark '3' means it does not meet the client satisfaction at all.

Results:
The table below shows the average rate of response to the above question.

Table of results – Adequacy of QA Procedures to Satisfy Clients' Requirements and Expectation

| | Key factors Adequacy Ratings | | |
	1	2	3
Overall Response			
1. Performance	84%	7%	9%
2. Features	53%	20%	27%
3. Reliability	59%	26%	15%
4. Responsiveness	9%	36%	55%
5. Competence	21%	44%	34%
6. Conformance	43%	33%	24%
7. Access	7%	14%	79%
8. Currency	18%	36%	46%
9. Courtesy	7%	14%	79%
10. Communication	12%	38%	50%
11. Perceived Quality	0%	39%	61%
12. Security	17%	39%	64%
13. Understanding	17%	51%	32%
14. Tangibles	26%	35%	39%
15. Cost	30%	45%	25%
AVERAGE RATINGS	27%	32%	41%

Comments:

From analysis of the results shown in the above table, the following picture emerges:

1   Taking an average of the ratings over all fifteen 'key factors' of quality, it appears that an average of only 27% of responses ranked the 'current' QA procedures as completely adequate to ensure that the clients' requirements and expectations are fully satisfied. 32% of the responses thought moderately and 41% thought that 'current' QA procedures do not cover the particular 'key factors' that were listed.

2   However, these somewhat negative results (relative to what might have been expected within a firm operating an accredited QA system) may be at least partially explained by reference to the hypothesis put forward earlier in this dissertation; that is 'a standard accredited QA system such as ISO 9001 may not sufficiently address all the pertinent aspects of quality'.

3   Going back to the above table, 84% of respondents ranked 'performance' (that is, ensuring that the required basic service is actually performed) as fully satisfying the clients' requirements with the 'current' QA procedures.

4   It was expected that reliability to be well within the remit of QA. Here, the results are a little lower: an average of 59% of respondents ranked this item as

fully satisfying clients' requirements, 26% ranked it as moderately and 15% as not at all. Further analysis of the results, into categories of responses, discloses that the majority of the low ranking came from the Construction Professionals which, presumably, are either genuinely critical to QA procedures or otherwise may lack knowledge as to the purposes and scope of the procedures in operation.

5   In a similar vein, the remainder of the overall 'top six' quality components appear to illustrate aspects of quality which are, indeed, less well covered by standard QA systems such as ISO 9001. The extreme example of this is 'perceived quality', which refers to the trustworthiness of the firm or contact personnel to the client.

6   Though proposed by the majority of respondents as an important aspect of quality in the selection, retention and recommendation of the professional firm's services, not one respondent suggested that 'current' QA procedures would ensure that clients' reasonable requirements and expectations are fully satisfied in this respect. Indeed, 61% indicated that they believed that 'current' QA procedures did not cover this factor. However, this finding should be treated with caution. The answers are the respondents' own perceptions and not necessarily the organisation's. A larger sample needs to be surveyed in order to confirm this finding.

7   Similar results apply to 'communication and responsiveness', where 50% and 55% of respondents, respectively, indicated that QA procedures did not cover these key factors.

8   The respondents were also asked to rank the 'five most important' factors of quality and the 'five least important' factors that the QA procedure should take into consideration. The following were the results:

MOST IMPORTANT FACTORS	LEAST IMPORTANT FACTORS
1. Reliability	1. Tangibles
2. Understanding	2. Features
3. Perceived Quality	3. Security
4. Performance	4. Currency
5. Communication	5. Conformance

Presenting postal survey supplemented by interviews (example of writing up style)

This example is related to health and safety in construction. Postal and interview questionnaires were designed. The postal questionnaires were sent to four companies, comprising of three contractors and one construction consultancy which were mainly QS and Project Managers. A total of 130 questionnaires were distributed of which 77 completed forms were returned (60 per cent response). The returned responses included 25 managers, 25 supervisors, 21 operatives and 6 consultants. All the returned questionnaires were usable. This high response to the survey was due to the fact that the author had a good contact base which was formed during the professional placement. To support the findings, the student also decided to undertake interviews with the companies who responded to the survey. The following section is an extract from the analysis of the results.

Question:
This question was related to Health and Safety as opposed to Programme Time. The respondents were, therefore, asked the following question:
Do programme times imposed through contract clauses directly affect safety?
The table below presents the results

Table of results – Safety culture and Programme time

Question	Managers Agree Disagree	Supervisors Agree Disagree	Operatives Agree Disagree
Do programme times imposed through contract clauses directly affect safety?	16 (65%) 9 (35%)	7 (28%) 18 (72%)	17 (80%) 4 (20%)

Comments:
The above table shows an interesting finding. It shows that, whilst senior managers and operatives agreed with the question, supervisors expressed a different view. Analysis of the results indicates that 65% of Senior Site Managers agreed that programme time imposed through contract clauses directly affect safety. It was commented that penalty clauses are too high and programme times are too tight, imposing a false induced money-consciousness in the industry. One manger stated during the interview:
'There is a lot of pressure brought about at the present time, clients want the shortest programme, the highest quality and lowest price. Therefore, there is an awful amount of pressure on everyone involved with the contract at site level, particularly the site managers. I agree programme times do have an effect on safety, its not a perfect world as we also put pressure on sub-contractors to get the work done, forcing them to carry out work in a dangerous manner.'
When supervisors were asked the same question, 72% disagreed with the question. During an interview with a supervisor, he stated:
'You can still produce to the required level and effectively if you plan safely into your production process. People tend to work better in a safer environment. Today we think more consciously about safety, so whilst there is a lot of pressure on the main contractor, the emphasis is on building it on time, with quality, within budget and in the safest manner.'

Another interviewee commented: 'Under current regulations and working practices, safety is a fatal element in any construction sequence, along with staffing,
materials and plant. It can not be differentiated or taken aside from. On that basis
you can not regulate it, in many ways it should be in the forefront.'

A third interviewee put the view that: 'You have to do it the right way, even
though it will cost more or take longer. This industry is very dangerous, a serious
accident will cost the contractor more than it would have cost to do it safely in the
first place. The trouble is that people do not see risks until it's too late.'

From the above analysis one can conclude that present contractual arrangements allow the pursuit of safety to come into conflict with the demand for the
cheapest and speediest completion of a contract. The widespread use of contract
penalty clauses and unrealistic productivity targets can result in a poor safety culture as it can place supervisors under considerable pressure to take the quickest
and cheapest option. However, the Health and Safety issue should be part of the
contractual package and, as one interviewee stated, it should be in the forefront.

Presenting results based purely on interviews (example of writing up style)

The table in the following extract shows the results of an interview with ten
clients who were asked questions that sought to compare management contracting with the traditional method. Structured interviews were used to sup-
plement data collected by questionnaire. The questionnaire was divided into
seven section headings to correspond with the research objectives. The following questions were investigated:

1   Who are the clients that used MC?
2   How do client organisations view MC?
3   Why have clients used MC?
4   What are the clients' criteria of satisfaction?
5   What is the level of client satisfaction with MC?
6   Is MC favourable to traditional contracting?
7   What are the clients' attitude towards the use of MC in the future?

In this example, the questionnaire was sent to ten client organisations prior to
the interview. The organisations were coded as client A, B, C, D, E, F, G, H, I, J.
Interviews were carried out by running through the questionnaire to comment
on the answers of each section heading, and were documented using a mini
tape-recorder. The following paragraphs provide you with a sample of a discussion that was made on the results related to question 6 above. This should give
you some idea of how to write and discuss the interview results.

Table of results – Responses to questions on MC versus traditional method

| | Response out of 10 | | |
	Yes	Same	No
1. Is MC more risky to clients?	6	2	2
2. Is MC more profitable to the contractor?	10	–	–
3. Does MC involve fewer claims?	3	4	3
4. Is MC more flexible to variations?	10	–	–
5. Does MC allow an earlier start on site?	10	–	–
6. Is MC quicker to design and build the project?	10	–	–
7. Is MC more reliable in predicting the build time?	9	1	–
8. Is MC more reliable in estimating the project cost?	6	3	1
9. Is MC cheaper?	2	4	4
10. Does MC provide more control for sub-contractors?	9	1	0
11. Does MC exercise more control over construction operations?	9	1	–
12. Does MC provide a better building design?	1	1	8

Comments:

1 As can be seen from the above table, there is a conflict of opinion concerning the risk to be absorbed by clients when dealing with a management contractor. Clients A, B and D saw the principle risk arising from the absence of a tendered lump sum price from the main contractor prior to construction. Client F claimed that:

'Clients are subject to a greater risk in respect to costs because of the staggering and phasing of orders for specific work over a longer period. While in the traditional method it was the main contractor who was taking that risk by putting a lump sum bid out at the outset. With MC the contractor is likely to settle for a smaller guaranteed profit and abandon a higher potential profit through the management of implicit risks. This balance between profit and risk has implications for clients. Larger clients may be better placed to take risks and consequently have a cheaper project.'

2   All clients interviewed agreed that MC is flexible in that it enables variations to the original design and specifications throughout the course of construction; they added that cost can be controlled by changes in the design without affecting project performance. Client H stated:

'with the amount of changes our organisation made for the last management contract, it would have been a disaster if we had used the traditional form of contract.'

3   However, not all clients reacted positively regarding the assertion that fewer claims arose with MC. Client D, while generally satisfied with MC, was unconvinced that MCs were less claims-conscious than traditional contracts. Clients A, B, F, H and J did not experience any differences in claims between traditional contracting and MC. This was unexpected because many articles that were published by management contractors stressed that the system involves fewer claims. In an article published by the *Building Magazine,* A management contractor stated that:

'you could run a MC without a form of contract. It is a philosophy, an attitude and a contract of trust.'

4   Frequently, the time factor was seen as one of the major advantages of MC: none of the client's sampled commented unfavourably about MC's time performance. All clients agreed that MC reduced the pre-contract period by overlapping the design and construction process. This enabled the project to be completed in a shorter time than for a traditional method. However, clients A, B and D expressed the view that their experience with past MCs counted for a good deal in considering the company's other needs.

5   Conflicting attitudes about the cost factor were noted. On interviewing a large public client (A), a mismatch between its expectations and the procedures of MC with its uncertain final costs, could be observed. Client (A) stated:

'Due to the way the public sector is organised, it is naturally biased towards caution in committing tax payers' money and ensuring that their accounting officer (i.e. Chief Executive) has answers to critical questions which might be put to him by the Public Accounts Committee.'

However, a second public organisation did not feel constrained in using management contracting due to public accountability because they had to change their building procedures. According to McKinney (1983), the public sector has experienced cost reimbursement contracts for many years. This particular form of contract would certainly seem to be far more in conflict with the concept of public accountability than would a 'pure' MC, simply because the contractor carries out some of the work himself with little provision for realistic and comprehensive competition. Traditionally, the public sector has taken a narrow view and awarded contracts by means of open competition on the basis of drawings and a bill of quantities.

A private banker (B) stated that 'there is a tendency for greater involvement of the professional consultants: the architect and quantity surveyor get involved more than they should in some work which is the MC's job. This overlapping responsibility was reflected in higher fees being paid.'

Four private clients had a fairly positive attitude toward the cost performance of MC. Clients C, E, F and G had a fairly positive attitude toward the cost performance of MC. One distinguishing characteristic amongst this group was that low costs were not considered as essential for client satisfaction.

6 None of the clients interviewed felt that MC produced a better building design than the traditional method, but most clients stated that they did not choose a MC for that reason in the first place.

7 Finally, the management contractors and clients who participated in the study criticised many contracting organisations for entering MC without the right personnel. Client C noted that, 'although from the client's point of view, the intention is to integrate them with the professional team of architect, structural engineers and quantity surveyors at an early stage in the proceedings to gain the advantage of their know-how within the building industry, many have not yet understood or chosen to understand this change in status and merely regard themselves as administrative middle-men in between the sub-contractors and client and the professional team and thus do not inject any creative ideas, which is one of the objects of the exercise and is indeed the reason why certainly in our case, after a careful selection process, we bring them into the proceedings at the earliest possible stage'.

Section 2: analysis of the results, part 2 (test of correlation and or association)

If you applied inferential statistical tests, this part of the dissertation will then consist of hypotheses testing, tables of correlation or association and interpretation of the results. The table in the following example shows an example of a survey extracted from an MSc dissertation investigating whether there are significant differences in opinions between head office personnel and site managers on factors that influence construction productivity. Several productivity factors were extracted from the literature review and were assessed by 19 head office personnel and 17 site managers. The findings were presented as follows.

Presenting results using inferential method of analysis (example of writing up style)

The research question	–	Do site managers and the head office perceive productivity factors differently?
The research hypothesis	–	There are significant differences in perceptions between head office personnel and site managers with regard to factors impairing productivity. These differences are related to management techniques, resource management, motivation and experience.
The null hypothesis	–	There is no difference in ranking to productivity factors between head office personnel and site managers.

Statement of hypothesis testing

The following table of results shows that there is a high correlation in the ranking between the two samples. Hence, the null hypothesis can be accepted concluding that head office personnel and site managers do not perceive productivity factors differently. The table also shows that there seems to be an overall agreement by both samples with regard to productivity factors (Rho = 0.78). This indicates that high ranking by head office personnel to the productivity factors correspond to high ranking by site managers to the same factors, and vice versa.

Comments:

Based on the above findings, the following interpretations are made:

1   The respondents identified 'ineffective project planning' as the most crucial factor likely to impair construction productivity. This finding supports Borcherding's notion in that 'the planning/design level is probably the key communication link in the hierarchy model between the "reality" of the site construction level and the "abstract" of the policy and programme levels. The planning/design level becomes the controlling element in the industry's effort to translate productivity information from above into a common language with a terminology meaningful to the desired audience' (Borcherding & Garner 1981).

The survey results – Analysis of the survey result is shown in the table below.

Productivity factors	Head office rank (A)	Site managers rank (B)	Difference in ranks $d_i =$ (A − B)	Difference in ranks square $d_i^2 =$ (A − B)^2
1. Head office factors Lack of integration of project information.	5	3	2	4

Ineffective project planning.	1	1	0	0
Poor scheduling of project.	8	11	−3	9
Under direction of employees.	10	9	1	1
Exclusion of site managers from contract meetings.	8	10	−2	4
Poor selection of project personnel.	11	12	−1	1
Lack of consultation in decision making.	8	13	−5	25
2. Resource management effectiveness				
Difficulties with the procurement of material.	5	3	2	4
Disruption of site program.	5	8	−3	9
Discrepancies with technical information.	10	6	4	16
Ignorance/lack of knowledge of project technology.	12	6	6	36
3. Motivational factors				
Low salary.	13	11	2	4
Constraints on a worker's performance (work env.)	2	3	−1	1
4. Experience and knowledge factors				
Lack of experience and training.	3	6	−3	9
Poor knowledge of scientific techniques (workers).	14	14	0	0

Total di^2 = 123

$$\text{rho} = 1 - \frac{6 \sum \text{di}^2}{N(N^2 - 1)} = 1 - \frac{6 \times 123^2}{15(225 - 1)} = 0.78$$

$P < 0.005$

Another highly scored factor within this section, was 'lack of inter-unit integration of project information', which is a factor related to the site-level construction information system. This emphasises the importance of focusing on information required to drive the building process in addition to the technical information for design-related problems.

The difference in perception between the two samples was registered in the area of 'lack of consultation in decision making'. It was identified that head office personnel assigned a moderate score to this factor while site managers gave it a low scoring.

2 The factors that seem to be of importance to both samples in questions related to resource management effectiveness are that of 'difficulties with the procurement of materials' and 'disruption of site programme'. As far as material mismanagement is concerned, comments from site managers indicated that most acute problems, caused by material mismanagement, are with material supply and storage, which had a great impact on the sequence of work, and rework due to disruptions. This finding add further support to Thomas's studies (Thomas et al., 1989, 1992) who found that the average daily productivity for the non-disrupted days was 0.44 work hours/sqm, while disrupted days had an average productivity of 2.16 work hours/sqm. This finding demonstrates the average degree of productivity improvement that can be achieved through better resource management control.

With regard to 'disruption of site programmes', this factor can cause delay in executing the work and prevent optimising utilisation of available resources. Timely input from all levels of management can reduce the risk of overlooking activities necessary to complete the project. Therefore, delays contributing to lowering a worker productivity can be reduced by planning the work to efficiently utilise manpower.

There was disagreement between both samples on 'discrepancies with technical information' and 'ignorance of project technology'. Head office personnel do not regard them as highly important, while site managers do. It was noted by site managers that waiting time due to discrepancies with technical information can be reduced by providing adequate supervision at the work place.

3 In reviewing the results described by previous researchers, there would appear to be a difference in views as to whether work force motivation can affect construction productivity. Several behavioural and physiological researchers indicated that the expenditure of effort by a worker is the physical manifestation of motivation; the greater a worker's motivation, the greater his/her expenditure of effort. Ranking on questions related to motivation show that there was little support to the factor of 'low salary'. However, a substantial proportion of both samples were in agreement with the statement that 'constraints on worker's performance arising from ineffective management will affect their motivation at work and consequently influences worker's productivity'. This finding is congruent with Borcherding's conclusions (Borcherding & Oglesby 1975; Borcherding & Garner 1981) in that the major sources of dissatisfaction commonly expressed, by both tradespeople and supervisors, are problems related to delay and reworking. According to Thomas et al. (1989), frequently cited problems include the lack of tools, materials, delayed decisions, late information and changes to orders. The link between satisfaction and improved productivity is based upon the impact that such conditions are assumed to have on workers' and supervisors' motivation. It can be concluded, therefore, that individual needs for opportunities for advancement as well as the

satisfaction from the work itself can be frustrating and, consequently, the potential for significant productivity gains is seen by the respondents as being thwarted by inappropriate or inadequate managerial actions.

4 Results to questions related to knowledge and experience show that both samples regarded 'lack of experience and training' as highly influential to site productivity. Obviously, each task requires specific skills and knowledge of how to use these skills. This finding ties closely with earlier results of this research regarding 'ineffective planning' and 'material mismanagement'. On the other hand, both samples do not regard 'scientific knowledge by workers' as crucial.

Writing the conclusion chapter

The conclusion chapter is the end product of your investigation and should not be confused with the results chapter or a summary. It should be tied very closely with your introduction chapter. Before writing the conclusion chapter, you need to read your research aim, objectives and hypothesis (or key questions), and start to conclude what you investigated within the broader context of your proposal. Start reading the whole dissertation thoroughly and make a note of key issues or findings. The contents of the conclusion chapter should include the following:

1 The dissertation aim.
2 Comments on the research objectives. Here take each objective in turn and conclude what you have achieved in each objective. Are your conclusions similar or dissimilar to previous writers in the subject area.
3 Comments on the research hypothesis or the research questions (whichever is applicable). Here state whether your investigation proved or disproved your research hypothesis.
4 Personal recommendations (practical) as to what should and should not be done. Perhaps a development of a conceptual model of a framework can be inserted in this part of the conclusion chapter.
5 Limitations of your research. Here state what were the difficulties that you faced throughout the research process which hindered you from achieving part of your objectives.
6 Further research.

Writing the references/bibliography

There are several different ways of citing references in your dissertation and describing the sources of your references, such as the Harvard system and the

Vancouver (numerical) system. Guides can be found in most libraries. It has to be stressed that it is essential that you use references in accordance with the guidance issued by your course, which may be included in unit guides or as a separate school publication. The following is a guide to issues about referencing that often cause some puzzlement. Before describing how to compile your references, general information on using the Harvard system in your dissertation is provided below.

What does 'referencing' mean?

References to the particular source are required if you draw upon another writer's ideas in your dissertation, or if you make a direct quote. Important reasons for referencing are:

1   To acknowledge other people's ideas.
2   To show that you are not just giving your own opinions but are also including those of other people.
3   To illustrate a point or to offer support for an argument you want to make.
4   To enable readers of your work to find the source material.

A well-used quotation can make a point clearly and concisely but try not to overuse quotations. It is important that you demonstrate your understanding of what you have read and are able to develop your own ideas by putting points or arguments into your own words.

How many references do I need to make?

There is no hard and fast rule as to how many references are required in a dissertation and much may depend upon the topic. Generally it would be impossible and unwieldy if you referenced every point you made but general guidance is that the main points or arguments you present, and especially direct quotes or ideas drawn directly from other people's work, should be acknowledged and referenced. Failure to give credit to other people's work appropriately could be considered as plagiarism.

What is plagiarism?

Plagiarism is to 'use the thoughts, writings, inventions, etc. of another person as one's own' (*Concise Oxford Dictionary*, 1995). Usually we are all formulating our own ideas from different information heard and read, but plagiarism involves a deliberate attempt to pass off someone else's ideas as your own, or to copy sections or chunks of texts from another's work without attempting to put these ideas into your own words or to acknowledge the sources. It is a form of theft and is certainly regarded as a serious offence in educational settings.

How do I incorporate references and quotations into my dissertation?

Using the Harvard system you must always give the author(s) and year of publication of your source. References can be incorporated into the text in a variety of ways. Some examples are:

1   In the following example you are making a statement in your own words and giving a reference for the source of your ideas or support for your argument:

> According to Haralambos and Holborn (1990), there are variations in the way deviancy can be defined in society today.

Further examples:

A recent television programme discussed … (BBC, 1994)
Research undertaken by Stainton Rogers (1991) indicated that …
Statistics show that … (Health lines, 1994)

2   In the following example you are making a direct quote (note that you must include the page number also):
> Haralambos and Holborn state that 'the family has been seen as a universal institute, an inevitable part of human society' (Haralambos and Holborn, 1990, p. 453).

3   With a longer quote it is generally preferred to indent the whole quote to make it clear:

> 'Even though attitudes of the clients are influenced by the experiences of individual organisations, there seems to be room for improvement in the management contracting system. One such shift would be for the construction industry, and in particular the management contractors, to adopt a professional – as opposed to a commercial – role. Yet this change may be shaped by clients, who can do much to fashion events by matching their own procedures to the requirements of the procurement method they have chosen. Project performance is often elusive, but the appropriate mix of client control and procurement method can make it less so.' (Shamil Naoum, 1994, p. 704)

How do I refer to a work which I have not seen myself but which is referred to in something that I have read?

These are known as secondary or second-hand references. Ideally they should be avoided as you should have read the original items yourself but, if it is unavoidable, it is important to make clear that this is a secondary reference, as follows:

> In an article of 1991 Wilford also maintains that Columbus's treatment of native people following his conquests is frequently seen in an ambiguous light (cited by Ransby, 1992).

The 'cited by' statement makes it clear that you have not seen the item by Wilford, only what Ransby says about it. Details of the item by Ransby will then appear in the references list at the end of your dissertation as this is the source of your information, i.e.

> RANSBY, B. (1992) Columbus and the making of historical myth. *Race and Class* 3(3) Jan–Mar, pp. 79–86.

Wilford must not appear in your reference list as you have not seen this article.

Compiling the reference list and bibliography

What is the difference between a reference list and a bibliography?

A reference list contains details of any item you have referred to or quoted from in your dissertation. A bibliography is a list of items you have made use of in preparing your dissertation, but not referred to in your text.

How do I set out my references list and bibliography?

Both lists are placed at the end of your dissertation and usually the reference list appears first. Both lists will be arranged in alphabetical order of author (following the Harvard system), and details given should be presented in a consistent format. Usually it is not necessary to include in your bibliography any item already listed in your reference list.

What details are required and where do I find these details?

Details required for a book can be found on the front and reverse of the title page which is usually the first or second unnumbered page inside the book.

Details for a book should be set out in the following order: Author, Initials (year) *Title* Edition (if later than first) Place of publication (Town): Name of publisher. Example:

> NAOUM, S. (1998) *Dissertation Research and Writing for Construction Students.* Oxford: Butterworth-Heinemann.

Details required for a journal article can usually be found on the contents list, front cover or the article itself. Details for a journal article should be set out in the following order:

Author, Initials (year) Title of article *Name of journal* Volume number (Issue number) Date of issue, First and last pages. Example:

> NAOUM, S. (1994) Critical Analysis of Time and Cost of Management and Traditional Contracts. *Journal of Construction Engrg. and Mgnt,* 120(4), Dec. pp. 687–705.

It is not necessary to put authors' names in capitals but it can help to make them stand out. Equally it can help readers if you underline titles or use a different typeface.

How do I give details from an edited book when the chapters are by different authors?

It is important to give details for each of the chapters you have referred to or used. The details of the chapter come first and then details of the book as a whole. Example:

> BREWSTER, C. (1992) Managing industrial relation in TOWERS, B. (ed.) *A Handbook of Industrial Relations Practice: Practice and law in the employment relationship* 3rd edn. London: Kogan Page. Ch. 8.

Points to note

1   *Authors:* If there are more than two authors it is acceptable to put the first author listed followed by et al. (which means 'and others') in the text. If you

need to refer to two or more items written by the same author in the same year use letters to differentiate the items, e.g. Smith (1994a), Smith (1994b).

2 *Dates:* Occasionally items do not give a date of publication in which case you put the abbreviation n.d. in brackets (this stands for 'no data'). This shows that you have checked for the data and not just forgotten to include it.

3 *Edition:* Only include edition statements if later than first. No edition statement indicates that it is a first edition. Make sure that you match the edition statement with the correct year of publication. This should be indicated clearly on the reverse of the title page.

4 *Page numbers:* You should always put a page number for a quote. In some cases the point you make might be referred to on a specific page or between specific pages (e.g. pp. 172–175). Sometimes it may be difficult to give specific page numbers as an argument may be supported by the item as a whole.

How do I describe a government report or an Act of Parliament?

Most reports will have as an author either a government department, body or committee, or may be well known by the name of the chairperson of the group or committee. With Acts of Parliament it is acceptable, and often easier, to cite the title of the Act in your text and then to include it in the reference list in alphabetical order of the first main word of the Act. Example:

SECRETARY OF STATE FOR HEALTH (1992) *Health of the nation: a strategy for health in England.* London: HMSO.

EDUCATION ACT 1988. London: HMSO.

NB HMSO is frequently the publisher of government reports and of all Acts. This abbreviation is acceptable. However, it is incorrect to describe HMSO as the author of a report, even if there is no clear author.

What about describing other types of material such as videos, leaflets, internet-based resources, etc.?

These are important sources of information but sometimes can be harder to describe. The golden rule always is to describe items as fully and clearly as possible, and in a consistent format. In the case of television/radio programmes make sure you note the date and channel of transmission. Unless self-evident it may be useful to indicate that an item is an audio- or video-cassette. Example:

CHANNEL 4 (1995) *Battered Britain: in search of law and order. 1: policing*, 8 Oct. London: Channel 4 (video-cassette).

In the case of leaflets, the author is often the organisation publishing the leaflet. Example:

London South Bank University (n.d.) *Referencing using the Harvard System: frequently asked questions.* London.

Finally, remember to:

1   Keep a careful note of all your sources as you prepare your dissertation.
2   Make a note of book details before you return them to the library.
3   Make sure you have the details you require on any photocopies you make.
4   Make sure you are referencing according to any guidelines set by your course.

This may seem like hard work but it is vital part of your dissertation and you may lose marks for incorrect referencing.

Summary

This chapter provided the guidelines that simplify the writing and organising of the dissertation material. It illustrated the logic of structuring the dissertation. In general, a dissertation is composed of three main parts. Part one covers the introduction chapter as well as two to three chapters of literature review. Part two covers the research design. Part three covers analysis of the results and conclusions. The main points to remember are:

1   The abstract should be brief, stating what you are investigating, how, and the main finding(s).
2   The introduction chapter should state the problem, purpose of study, aim, objectives, hypothesis and contents of the dissertation.
3   The literature chapters should be divided into sections with specific themes. All quotations must be acknowledged. At the end of each chapter, you need to appraise (critically or otherwise) the material that you included.

4   The research design chapter should give the technique that you used to gather the information, the rationale to the questionnaire, the source of data, characteristics of the sample, the method of analysis and/or format for presenting the findings.

5   The results chapter should clearly state the issues under investigation, the findings, and comments or interpretation of the results.

6   The conclusion chapter should comment on the objectives and hypothesis(s) that are stated in the introduction chapter.

7   The references/bibliography should follow a uniform system.

8   Other mechanics of presenting a dissertation, such as length and page formatting should be followed in accordance with the rules and regulations set by your institution.

References and additional reading

American Psychological Association (1974) Publication Manual. *Journal of Personality and Social Psychology*, **29**, pp. 80–85.

Bell, J. (1996) *Doing Your Research Project: A Guide for First Time Researchers in Education and Social Science*. Open University Press.

Cooper, H. (1995) Quality Assurance in Practice. BSc dissertation, School of Construction, London South Bank University.

Green, S. (1996) An Investigation into the Effect of Safety Culture and Site Operatives Behaviour on Construction Site Safety in the UK. BSc dissertation, School of Construction, London South Bank University.

Haywood, P. and Wragg, W. (1982) *Evaluating the Literature,* Rediguide 2, University of Nottingham, School of Education.

Howard, M. (1995) Marketing and the Quantity Surveyor: Understanding clients. MSc dissertation, School of Construction, London South Bank University.

Naoum, S. (1989) *An Investigation into the Performance of Management Contracts and the Traditional Methods of Building Procurement.* PhD thesis. Brunel University, Middlesex.

Naoum, S. (1991) *Procurement and Project Performance: A Comparison of Management and Traditional Contracting.* Occasional Paper No. 45, Chartered Institute of Building.

Naoum, S. and Hackman, J. (1996) Do Site Managers and the Head Office Perceive Productivity Factors Differently? *Journal of Engineering, Construction and Architectural Management*, **3**, March/June, pp. 147–159.

Naoum, S. and Langford, D. (1987) Management Contracting: The Clients' View. *Journal of Construction Engineering and Management*, American Society of Civil Engineers, **111** (3), Sept., pp. 369–384.

South Bank University (n.d.) *Referencing Using the Harvard System: Frequently Asked Questions.* London South Bank University.

10 *Dissertation supervision and assessment*

Once you have decided on the topic of your study and your proposal gets approved, your department should then appoint a personal supervisor for you. It is therefore important to know the role of your supervisor and what is expected from him/her.

This chapter will discuss the basics of what you should expect from your supervisor. It first explains what the role of the supervisor is and what it is not. Second, it illustrates how you may plan your dissertation and finally how would you expect to be assessed in the end.

It has to be stressed, however, that the information provided in this chapter is what is known as the standard and different institutions might have other requirements and regulations.

Getting started

Most universities run dissertation workshops that cover a range of issues such as: identification of research questions, assessing whether the topic is researchable, planning and undertaking a literature review, research strategies, research techniques, data analysis, writing and structuring the dissertation.

Following these workshops, each student will then submit a dissertation proposal (similar to the ones shown in Appendix 1). This proposal usually gets approved by the dissertation tutor (a person who is responsible for the dissertation module in your department).

After that, the dissertation tutor will allocate you a supervisor to look after your work. This allocation takes place during your study term and usually in the middle of the last term.

Once you are allocated a supervisor, it is your responsibility to arrange meetings with him/her. It is advisable to arrange a meeting with your supervisor as early as possible. Your supervisor should have a copy of the approved proposal, which forms the basis of discussion in these initial meetings.

It is strongly recommended that students and supervisors should meet at least twice following their initial meetings (proposal stage): once in the middle of the process to discuss progress, and once in the final stages to discuss the final draft. Additional meetings may be required for some students to discuss progress and drafts of individual chapters as they are written (see Dissertation supervision log section on p. 165). These additional contacts may take place electronically, for example by e-mail.

Note to students: It is unlikely that you can change your allocated supervisor. On rare occasions, students may find it difficult to work with the allocated supervisor. In such cases, the dissertation tutor should act as a mediator to resolve the differences through some agreed action plan. If this fails, then an alternative allocation should be arranged. It must be stressed, however, that if this unlikely situation occurs, it should be acknowledged and sorted as soon as possible before it is too late. Your supervisor is a resource, so use him/her to your advantage. He/she has normally been allocated time by the department to supervise you, so make sure you take advantage of this, but bear in mind that your supervisior is there to help you, not to do your work for you.

The role of the supervisor

The role of supervisor is to guide students towards the production of their dissertation by discussing each part of the process. The role of your supervisor can be summarised as follows:

1   To help you to develop your idea/topic further.
2   To help you to focus on your aim and objectives.
3   To help you sharpen your research questions and/or hypotheses.
4   To provide you with guidance about the appropriate research approach.
5   To approve your questionnaire before you distribute it to respondents.
6   To assist you in structuring your dissertation.
7   To discuss and monitor your progress.
8   To provide you with feedback on your chapters. This may be in the form of criticisms of your work.
9   To help you to present the final dissertation.

What is not the role of the supervisor?

From the above section, it is clear that your supervisor's duty is to guide you so that you can produce your best work. However, there are certain limitations as to what is expected from your supervisor. These can be summarised as follows:

1 Supervisors are not obliged to act as proofreaders of your work.
2 Students should not expect their supervisor to provide detailed feedback on more than one draft of each chapter (writing a dissertation is not an iterative process), or to correct spelling, grammar, punctuation, etc.
3 It is recommended that supervisors read the whole of the final draft before submission. However, if they do so, it is on the understanding that the result of the final examination is not in any way pre-judged.
4 Supervisors are not to assist with continual revision until the dissertation has acquired a certain grade that you may have as a target. Therefore, the supervisor's approval of your progress cannot be taken to imply any particular grade or classification. It is not expected that you request this from your supervisor at any stage of the dissertation module.

Planning your dissertation

In order not to let things drift, it is absolutely crucial to plan and time-manage your dissertation process. Figure 10.1 is an example of a work schedule for a typical dissertation.

Note to students:

1 Many of the above activities may overlap or take place before or after the suggested activities within the work schedule. For example, writing your literature chapters may take place while analysing your questionnaire.
2 There might be other activities involved in your research. For example, your dissertation might include a separate chapter on 'Discussion'. Or, you might wish to include a separate chapter titled 'Critical appraisal of literature' and insert it after Chapter 3 in the suggested work schedule shown in Figure 10.1. Your dissertation topic may necessitate a further chapter of literature review, so you should allow time for it.

Therefore, you need to design your work schedule to fit your own needs.

Activity	Planned starting date	Tick when completed	Comments
1. Identify a topic			
2. Undertake background reading			
3. Identify working title			
4. Prepare dissertation proposal			
5. Proposal approved by dissertation tutor			
6. First meeting with your supervisor after proposal approved			
7. Undertake literature review			
8. Building up your literature file			
9. Designing a questionnaire (if applicable)			
10. Designing a data collection format for secondary data (if applicable)			
11. Questionnaire or data collection format approved by your supervisor			
12. Building up a list of respondents for postal survey or interviews			
13. Sending out questionnaire			
14. Conducting interviews (if applicable)			
15. Collating secondary data for analysis (if applicable)			
16. Analysing the questionnaire, interviews or secondary data			
17. Write Chapter 1 – Introduction			
18. Write Chapter 2 – Literature			
19. Write Chapter 3 – Literature			
20. Write Chapter 4 – Research design and methodology			
21. Write Chapter 5 – Findings			
22. Write Chapter 6 – Conclusions			
23. Write list of references			
24. Write abstract			
25. Give final draft to a critical reader to proofread			
26. Give final draft to your supervisor for final comments and feedback			
27. Write acknowledgement			
28. Insert appendices			
29. Insert list of tables and figures			
30. Send to binding			
31. Submit			

Figure 10.1   *Example of a typical dissertation work schedule*

Dissertation supervision log

As mentioned above, you should arrange 'formal' meetings with your supervisor at least three times throughout the whole process. I call them 'formal' because, at each of these meetings, it is recommended that you fill out a 'log' form like the one shown in Figure 10.2. Some institutions even require these logs to be signed by the supervisor.

The purpose of these log forms is to keep a record of your progress and to act as milestones. You can arrange as many meetings as you and your supervisor think necessary, but the three essential meetings are:

Meeting 1:   Proposal stage
Meeting 2:   Progress
Meeting 3:   Final draft

Note to students:

1   You may need more than one progress meeting.
2   After you complete the log form it is strongly recommended that you send a copy to your supervisor for information and filing. You can do this by electronic means.

Dissertation assessment: Qualities in assessing at MSc level

In the following notes an overview is given about the qualities expected in the three marking bands (distinction, pass and fail) at London South Bank University Buckley and Naoum (2005). I cannot stress enough that this is only an example of assessing the quality of dissertations at MSc level and it may not be considered the norm. Other universities or courses, such as BSc, may apply different qualities of assessment.

There are three ways of considering the qualities of a masters dissertation:

(1)   descriptions of qualities;
(2)   judgement of the degree of difficulty; and
(3)   key words. They are designed as prompts for the marking teams and the qualities should be considered in combination and should not be interpreted mechanically.

1.   Descriptions of qualities

Distinction (70 marks or over)

The dissertation will contain some originality in the research subject chosen or in the research questions that have been developed. The research will show

---

Dissertation Supervision Log

Student's Name _____   Supervisor's Name _____

Meeting Number _____   Date:_____

Issues Discussed/Comments:

1.

2.

3.

4.

5.

Figure 10.2   *Example of a typical log form*

```
Actions Required:

1.



2.



3.



4.



5.
```

Figure 10.2   (*Continued*)

strong evidence in at least one of the degrees shown in (2) below. The study will demonstrate a sustained and rigorous use of theoretical/conceptual or empirical evidence to support the argument or discussion. Definitions and assumptions will be clear throughout the work. There will be strong evidence of critical analysis, judgement, originality or interpretation.

There will be discussion of methods and sources that link the research design or research approach to an appropriate literature about research methods. There will be reflection about the research process or the questions for further research, which are suggested by the findings of the study. There must be an argument put forward to support the methods adopted.

The study will be well structured and written in good English. A coherent and well-considered argument will be developed. Accurate referencing of sources will be shown in an appropriate way throughout the study.

Pass (50–69 marks)

The study will include a systematic review of relevant literature or other sources of information or data. There may be some evidence of originality in the research question(s) or in the approach taken to the subject. The research will contain a synthesis of appropriate ideas, evidence or data. Good use will be made of data and information to highlight new knowledge, to support argument, or to criticise current ideas, approaches, research methods or practices.

The discussion of methods and sources will include arguments in support of the methods or approach taken, or a description of the project design. There will be some references to the research methods' literature or similar appropriate material. An attempt will be made to reflect on the research method used.

The dissertation will be well structured and written in good English. Accurate referencing of sources will be shown in the appropriate way.

Fail (49 marks or under)

The research question(s) will not be well developed. There will be a limited review of appropriate literature or sources of information or data. The dissertation will contain an incomplete or sketchy treatment of theoretical or empirical material. The development of the data or information collected will be descriptive and may not be used in the right way. Argument is less likely to be developed or sustained.

Discussion of research methods will not be developed and is likely to contain important errors. There will be a lack of judgement about the research method used.

The dissertation will be poorly written. Referencing will be generally inaccurate or inadequate.

2.   *Degree of difficulty*

The degree of difficulty of the research attempted needs to be distinguished. This can be done in combination with the description (1 above) and the key words (3 below) by judgement of **one or more** of the following four criteria:

(a)   the degree of difficulty contained within the research question(s);
(b)   the degree of difficulty contained in the methodology;
(c)   the degree of difficulty in synthesis and the organisation of the research findings;
(d)   the degree of difficulty in the evidence collected or arguments presented.

3. *Key words*

Distinction	Pass	Fail
accuracy	analysis	brief
ambition	application	description
appraisal	calculation	error
balance	demonstration	incomplete
critical	experiment	irrelevant
evaluation	ideas	narrow
evidence	knowledge	omission
imagination	organisation	superficiality
interpretation	reflection	vagueness
judgement	review	
originality	sources	
rigorous	structure	
scope	synthesis	
style		
testing		
theory		

Note to students: Distinction level should also include criteria of the Pass category in (3) above.

Summary

This chapter explained the role of the supervisor and the importance of planning your research to achieve the desired outcome. In short, your supervisor will provide you with the appropriate direction towards the production of your dissertation. He/she can advise you on the research strategy, methodology and structure of your dissertation. Your supervisor is not obliged to proofread or correct spelling and grammar.

It is essential that you time manage your work. Therefore, drawing up a realistic work schedule is highly recommended.

It is also important to keep in close contact with your supervisor. You should arrange a formal meeting with him/her at least three times throughout the whole process.

Finally, it is also recommended to keep the log form up to date in order to keep record of your progress.

Reference

Buckley, M. and Naoum, S. (2005) *Dissertation Guide: Instructions for Preparation, Control and Presentation of the Dissertation*. Department of Property, Surveying and Construction, London South Bank University.

Appendix 1
Examples of dissertation proposals

Proposal 1: proposal for a descriptive type of research

Working title

Marketing and the private house-building companies: which marketing philosophy do companies have?

Rationale

The subject of this proposal developed from a personal interest in the dichotomy between technical excellence and customer orientation. In their book *In Search of Excellence*, Peters and Waterman[1] point out that the most successful US firms were characterized by a common dedication to marketing. More specifically the key to success lies in 'keeping close to the customer'.

The size of private housing market in relation to total construction output is rather significant. It accounts for 14 per cent of construction output for an all work total and 24 per cent of construction output for total new work.[2] Hence, the proposed study is an important one and the author will be focusing on the marketing philosophy of the private house-building companies.

Literature provides numerous alternative definitions of marketing; most have certain basic features in common, especially the notion of looking at the firm from the point of view of the customer or striving to ensure mutual profitability from the marketing exchange. Other definitions place the emphasis on the essential managerial nature of marketing. Marketing, however, should be distinguished from the selling.[3] Selling is only one of several marketing functions and may not be the most important.[4] In addition, selling may require a different set of skills from those needed for marketing.[5]

The principal theme of this research is to use Stanton and Futrell's [6] marketing profiles and place each of the companies to be studied in a particular category. Stanton and Futrell established three marketing profile categories, namely, Production Orientation, Sales Orientation and Marketing Orientation.

The following sections of this proposal give further details about the principal objectives for this research.

Aim

To establish the marketing philosophy of private house-building companies and to relate it to the financial performance of each company.

Objectives

1   To survey large private house-building companies and investigate the following:
    (a)   their business philosophy
    (b)   organisational structure
    (c)   marketing information system
    (d)   marketing planning and policy.
2   To test the relationship between the companies' marketing philosophy and their financial performance.

Hypothesis

1   Most private house-building companies adopt the 'sales oriented' philosophy.
2   Private house-building companies that adopt a 'high marketing profile' perform better financially.

Methodology

Stage 1: literature review

This study will be confined to the private house-building sector where there is a more direct link between the company and the end user. The initial survey of the literature reveals no specific study of marketing in the private house-building industry. However, the desk work for this study will concentrate on research which has looked at the construction industry in general, such as, the earlier research of O'Callaghan[7] and Stilgoe.[8] This should provide a useful point of reference.

Stage 2: pilot study

Data collection will take the form of a structured postal questionnaire. However, an initial pilot study will be conducted to test the validity of the questionnaire through in-depth interviews with marketing managers.

Stage 3: main survey questionnaire

The feedback from the pilot study should assist in finalising the questionnaire and prepare the ground for the main survey. The questions will centre around the areas mentioned in the above objectives. In order to obtain a high level of response, a multi-option format will be designed and limited to about 30 questions. Open questions are considered to be inappropriate as they would require the respondents to formulate an answer which needs lengthier input and will therefore be more difficult to analyse.

Stage 4: analysing the postal questionnaire

The analysis of the questionnaire will take two forms. First, to determine the level of agreement or disagreement by the respondents to each question within the questionnaire by counting the number of respondents who answered favourably or unfavourably. Judging by the response, each company will be placed in one of the three marketing profile categories, i.e. Production Orientation, Sales Orientation or Marketing Orientation. Second, to correlate each of the marketing philosophy questions to the financial performance of each company. Financial information will be obtained from the business ratio report and from companies' annual reports. The level of financial performance will be gauged by establishing a marketing rating system applicable to private house-building companies.

Stage 5: writing the research report

This stage involves writing up the content of the dissertation and should cover the chapters proposed in the following section.

Proposed structure of the dissertation

Chapter 1 – Introduction.
Chapter 2 – Marketing theory and principles.
Chapter 3 – Marketing and the construction industry.
Chapter 4 – Research design and methodology.
Chapter 5 – Analysis of the results and discussion.
Chapter 6 – Conclusions.
Chapter 7 – Suggestions for further research.
Chapter 8 – References.

Initial references

1   Peters, T.Z. and Waterman, R.H. (1982) *In Search of Excellence: Lessons from America's Best Run Companies*. New York: Harper and Row.
2   National Economic Development Office (1988) *Construction Forecast 1988– 1989–1990*. London: HMSO.
3   Foster, D. (1984) *Mastering Marketing*. Hong Kong: Macmillan.
4   Kotler, P. (1986) *Principles of Marketing*. Third edition, London: Prentice-Hall.
5   Fisher, N. (1986) *Marketing for the Construction Industry*. London: Longman.
6   Stanton, W. and Futrell, C. (1986) *Fundamentals of Marketing*. International edition. New York: McGraw-Hill.
7   O'Callaghan, J. (1986) Managerial Characteristics and Financial Performance of Construction Companies. MSc thesis, Brunel University.
8   Stilgoe, G. (1985) Marketing Policy and Construction Firms. MSc dissertation, Brunel University.

Programme of work

In this section of the proposal, a bar chart should be attached showing the programme of work for the dissertation. It should also indicate the target dates for the completion of each of the stages shown in the methodology section.

Proposal 2: proposal for the analytical approach/theoretical framework

Working title

A critical analysis of the relationship between the building team, the procurement method and the project performance.

Rationale

It is axiomatic of construction management that a project may be regarded as successful if the building is delivered at the right time, at the appropriate price and quality standard as well as achieving a high level of client satisfaction. Increasingly, the achievement of these criteria has been associated with the method of procurement. In short, the selection of the appropriate method can shape the success of the project and this research has chosen to study the performance of management contracting and traditional contracting.

The problems associated with the building process have been considered in relation to separation of design from construction[1,2] lack of integration,[3] poor communication[4] and increasing project complexity.[5] These factors, together with economic changes, e.g. inflation and recession,[6] have led construction professionals and the industry to offer alternative methods of procurement, such as management contracting, design and build and project management.

With the growth in the use of these methods in the mid-1970s, a number of researchers have investigated the management of the building process and in particular the performance in terms of time, cost and quality. In one study, the relationship between contractual arrangement and project success was investigated on 32 building projects, from which it was found that the essential element which brought about success was the level of managerial control.[7] Of the contractual methods studied, those with a high level of managerial control, e.g. management contracting, performed better on time, gave a high level of client satisfaction and overran the budget by less.

The case study approach has been used previously to illustrate the performance of management contracting, but the results were based either on small samples or the number of management contracts studied were very limited. Moreover, there have been relatively few investigations which have analysed the characteristics of the client, the designer and project within the scope of one study.

Aim

To use the contingency theory to develop a conceptual model which can be used to analyse the relationship between project performance and procurement methods of commercial buildings.

Objectives

1 To establish the background, apparent advantages and disadvantages to the client in adopting management and traditional contracting.
2 To evaluate the client criteria and priority when selecting management contracting.
3 To find out whether the characteristics of the client have a significant effect on project performance.
4 To find out whether the characteristics of the designer have a significant effect on project performance.
5 To find out whether the characteristics of the project have a significant effect on project performance.
6 To compare the performance of time and cost of commercial projects completed under management and traditional contracts.

Theoretical framework

The theory that I will use will be the contingency theory. It was developed by Kast and Rosenzweig[8] and it was used to study the management of organisation, including organisation design, leadership, behavioural change and operations. The contingency theory indicates that there is no one way to design, manage or to organise a business enterprise, it is contingent or dependent upon the problem or situation in hand. As applied to my study, this theory will hold that I would expect my independent variables (client and project characteristics), to influence the intervening variable (procurement method) and in turn influence the dependent variable (project performance). This proposition is thought to apply to my research because the contingency approach recognises the complexity involved in managing modern organisations, but uses the pattern of relationships and/or configurations of subsystems in order to facilitate improved practice.

Main hypothesis

In order to achieve a high level of project performance, the procurement method should be selected in accordance to the characteristics of the client, designer and the project.

Outline methodology

Stage 1: literature review and the pilot study

This study will review the relevant literature on the subject of procurement methods, in particular, looking at previous building process models which studied the performance of the procurement method. The literature review will be followed by a pilot study which will take the form of structured interviews with two clients and two contractors who have commissioned and experienced 'pure' management projects. An interview questionnaire will be used for this pilot study in order to validate the appropriateness of the main study questionnaire.

Stage 2: the main study

The pilot study should prepare the ground for designing the main study questionnaire which will be used to compare the performance of management and traditional contracts. It is intended to adopt the quantitative data collection approach, which means obtaining facts and figures from previously completed projects and not solely taking opinions or views of the respondents. It is also intended to obtain a large sample of cases, approximately 60 projects, all of which will be analysed statistically.

Stage 3: writing up

This stage involves writing up the content of the dissertation and should cover the chapters proposed in the following section.

Proposed structure of dissertation

Chapter 1 – Introduction and background to the problem.
Chapter 2 – Review of the traditional procurement system.
Chapter 3 – Review of management contracting.
Chapter 4 – Research design and method of analysis.
Chapter 5 – Analysis of the results.
Chapter 6 – Conclusions.
Chapter 7 – Bibliography.

Initial references

1   Emmerson, H. (1974) *Survey of Problems before the Construction Industries.* London: HMSO.

2   Banwell, H. (1964) *The Placing and Management of Building Contracts.* Ministry of Public Building and Work, London: HMSO.
3   Sidwell, A. (1979) *A Field Study of Organizational Forms.* Birmingham: Department of Construction and Environmental Health, University of Aston.
4   Higgin, G. and Jessop, N. (1965) *Communication in the Building Industry.* London: Tavistock Institute.
5   Bennett, J. and Fine, B. (1980) *Measurement of Complexity in Construction Projects.* Reading: Department of Construction Management, University of Reading.
6   Hillebrandt, P. (1977) *Economic Theory and the Construction Industry.* London: Macmillan.
7   Sidwell, A. (1982) 'A Critical Study of Project Team Organizational Forms within the Building Process. PhD thesis, Department of Construction and Environmental Health, University of Aston.
8   Kast, F. and Rosenzweig, V. (1973) *Contingency Views of Organization and Management.* London: Science Research Associates.

Research programme

In this section of the proposal, a bar chart should be attached showing the programme of work for the dissertation. It should also indicate the target dates for the completion of each of the stages shown in the methodology section.

Proposal 3: proposal for the problem-solving approach

Working title

An evaluation of the total quality management system: a case study.

Rationale

The problem of quality and its associated costs has been of great concern to most industries since the early 1960s. As a result, some large construction contractors have pursued and implemented innovative quality management techniques such as quality assurance (QA) and total quality management (TQM). Oakland (1990) defined TQM as a way of managing to improve the effectiveness, flexibility and competitiveness of a business as a whole. It applies just as much to service industries as it does to manufacturing. It involves whole companies getting organised in every department, every activity and every person at every level.

Once management acknowledges that there is a problem, it can take the second step to develop a clear understanding of underlying principles and elements of TQM such as training (Imai and Kaisen, 1986), team work (Aubrey and Felkins, 1988), supplier involvement, customer services and its implementation which is considered by Juran (1988) as the primary quality management tool.

According to Boardman and Croxson (1994), several senior TQM representatives from various client and contracting organisations forming the European Construction Institute TQM Task Force, were discussing this very topic of measuring where their companies were on the TQM journey and it became manifestly obvious there were various methods for the measurement of TQM. The preliminary literature review, revealed that Boardman's measurement chart is simple but comprehensive and can be used to give a visual representation of the organisation's current position in a TQM programme.

Based on such a background, this research will be using the matrix model developed by the Trafalgar House Quality Management Services organisation as the tool to identify the specific problems within the author's organisation and to make recommendations, if appropriate, for improving the quality system.

Aim

To analyse the present quality system in the organisation and to make proposals for possible future action relating to the introduction of a total quality management system.

Objectives

1   To review the concept and principle of TQM.
2   To assess the advantages of the TQM systems.
3   To measure the efficiency of the current quality management system in the author's organisation and to identify any present problems within the system.
4   If appropriate, to design and recommend a new TQM system for the organisation.
5   To evaluate the proposed TQM.

Hypothesis

Note to students: With the problem-solving approach, formulating a hypothesis is not usually applicable. It is only applicable when the proposed solution is to be put under test, which is not practical for the level of undergraduate student. Instead, a number of key questions have been formed.

Key questions

1   What are the reasons that drive the construction industry towards accreditation of standard quality assurance?
2   Does accreditation of QA assure the customer to an optimum standard of services?
3   What are the differences between, quality assurance and total quality management?

Methodology

Stage 1: literature research

A comprehensive review of the relevant literature including a computer-assisted search will be undertaken in order to develop an understanding of previous work in the field of TQM.

Stage 2: exploratory interviews

This stage will explore the matrix designed by the Trafalgar House Quality Management Services House. The exploration will be achieved through unstructured interviews with five companies that are known to have implemented a TQM system.

Stage 3: case study

The author will then use his or her own organisation as a case to study the possible implementation of a systematic TQM. Interviews with employees will take place to score the level of quality effectiveness.

Stage 4: evaluation of the organisation's present TQM system

This stage will bring together and review the information collected in the previous stages in which the present TQM system has to be studied.

Stage 5: writing up

This stage involves writing up the content of the dissertation and should cover the chapters proposed in the following section.

Proposed contents of dissertation

Chapter 1 – Introduction.
Chapter 2 – Total quality management – the concept.
Chapter 3 – Implementation of TQM.
Chapter 4 – Designing the case study and research methodology.
Chapter 5 – Analysing and evaluating the case study evidence.
Chapter 6 – Conclusions and further studies.
Chapter 7 – References.

Initial references

Aubrey, C. and Felkins, P. (1988) *Teamwork: Involving People in Quality and Productivity Improvement.* Milwaukee: American Society of Quality Control.
Boardman, R. and Croxson, J. (1994) Trafalgar Quality Management Services. Unpublished report by the Trafalgar House Company, UK, 25 pages.
Imai, M. and Kaisen, L. (1986) The Key to Japan's Competitive Success. New York: Random House.

Juran, J. (1988) *Juran on Planning for Quality*. New York: The Free Press.
Oakland, J. (1990) *Total Quality Management*. Oxford: Butterworth-Heinemann.

Programme of work

In this section of the proposal, a bar chart should be attached showing the programme of work for the dissertation. It should also indicate the target dates for the completion of each of the stages shown in the methodology section.

Proposal 4: proposal for the desk study approach – analysis of secondary source of data

Working title

Safety and the construction industry: a statistical analysis of facts and figures of the construction industry accidents.

Rationale

In the construction industry, the risk of a fatality is five times more likely than in a manufacturing-based industry, whilst the risk of a major injury is two and a half times higher (Davies and Tomasin, 1990). A Health and Safety Executive (HSE, 1983) study, reveals that construction has one of the highest ratios of non-injury to injury accidents of all UK industries. For every injury on site, the HSE recorded 3570 situations that could have led to injury (*Construction News*, 1994). Moreover, the following table shows that each year there are over 120 people killed and over 3000 suffer major injuries in construction related accidents. Include the 17,000 other injuries reported to the HSE and the estimate of under-reporting of injuries of between 20 and 50 per cent, and you can see that the accidents rate in the construction industry is unacceptable (Barnard, 1995).

Accident Figures 1990/91: Health and Safety Executive Annual Report

Accident type	Employees	Self-employed	Public	Total
Fatal	93	27	9	129
Major	2,894	936	112	3,942
Over 3 days	16,390	1,539	–	17,929
All	19,377	2,502	121	22,000

Apart from the human cost of suffering an accident; the economic effect can be devastating. Direct costs of accidents such as injury, illness or damage can be insured against, although this is the tip of a very large iceberg. Every £1 of an accident costs that an insurance company has to pay out, could cost the contractor between £5 to £50 in indirect costs. These indirect costs will range from product and material damage, to legal costs (Travers Morgan Ltd, 1995).

Aim

To analyse accident statistics within the construction industry and to show whether the economic climate has an effect on the rate of accidents in the industry.

Objectives

1   To provide a general overview of the history of accidents in the construction industry.
2   To compare the accident rate of the construction industry with other major industries within the UK.
3   To show the trades that have the highest accident rate by analysing the ratio of accidents to employees.
4   To examine the cost implications of construction site accidents.
5   To find out whether there is a correlation between the state of the general economy and the rate of accidents.

Key questions

1   What is the ratio of accident/number of employees in the UK construction industry?
2   Does the recession have an impact on site safety records?
3   What is the cost implication of accidents in the construction industry?

Methodology

Stage 1: literature review

The basic concern throughout the review stage is to identify some of the broader parameters likely to be relevant in studying the construction industry's safety statistics. In order to achieve the first objective, a systematic literature review is to be conducted which will cover textbooks, institutional and statutory publications, periodicals and trade/academic journals, seminar and conference papers.

Stage 2: analysing the facts and figures

The second and third objectives of the study can be achieved by analysing published and unpublished accident statistical data which will be acquired from the Health and Safety Statistical Services Department in Liverpool.

The fourth objective will be achieved by gathering information from the Health and Safety Executive Report, entitled, *Cost Benefits of Implementing the CDM Regulations 1994*. This objective will show how the distribution of accident costs affect the industry and society.

To achieve the fifth objective, information about the economic climate during the last decade will be gathered from the HMSO construction output and then correlated with the accident record, in an attempt to find out whether the rate of accidents is related to recession. The statistical package SPSS will be utilised to assist in evaluating this relationship.

Stage 3: producing tables and graphs to present the results

Note to students: This is part of the writing up stage where all the results will be presented in graphical format to allow discussion and interpretation to be made on the research findings.

Proposed structure of dissertation

Chapter 1 – Introduction to the safety problem.
Chapter 2 – Health and Safety and the employment trend.
Chapter 3 – Safety culture in the construction industry.
Chapter 4 – Previous empirical research on construction safety.
Chapter 5 – Analysis of the secondary data.
Chapter 6 – Discussion and critical appraisal.
Chapter 7 – Conclusion and recommendation for further research.
Chapter 8 – References.

Initial references

Barnard, M. (1995) Safety is the Engineers Responsibility: Facts or Fiction? A paper given by the Director of Health and Safety for Travers Morgan Ltd at the Institute of Civil Engineers on Construction Site Safety and the CDM regulations, 6 February, School of Business and Industrial Management.

Construction News (1994) Preparing for Safety Laws. *Construction News*, 1 December, p. 16.

Davies, V. and Tomasin, K. (1990) *Construction Site Safety*, London: Thomas Telford.

HSE (Health and Safety Executive) (1983) *The Costs of Accidents at Work*. London: Health and Safety Commission, HMSO.

Travers Morgan Ltd (1995) The Economics of Safety Management. A paper given by Mr Mike Crocker of Travers Morgan Ltd at Walfords, London.

Proposed programme of work

In this section of the proposal, a bar chart should be attached showing the programme of work for the dissertation. It should also indicate the target dates for the completion of each of the stages shown in the methodology section.

Appendix 2
Construction management and economics list of journals

Produced by Dr Farzad Khosorowshai (editor ARCOM Newsletter)
ARCOM Newsletter; vol 10, no 3, August 1995.

All England Law Reports
Appropriate Technology
Architect, Builder, Contractor and Developer
Architect and Surveyor (now *Building Engineer*)
Architects Journal
Asian Pacific Building and Construction Management Journal
Bank of England Quarterly Bulletin
Banking World
BCIS Quarterly Review of Building Prices
BMI News
BMI Publications Digest
BMI Quarterly Cost Briefing (*Building Maintenance Costs*)
BRE News of Construction Research
British Economy Survey
BSRIA Statistics Bulletin
Building
Building and Environment
Building Control
Building Design
Building Economist (previously *Chartered Quantity Surveyor*)
Building Economist (*Australia*)
Building Engineer (previously *Architect and Surveyor*)
Building Law Reports

Building Market Report
Building Research and Information
Building Services (Subzero Journal)
Building Services and Environmental Engineer
Built Environment
Bundesbaublatt
Business Briefing
Business Monitor MQ5 – Quarterly Statistics
CAD/CAM International
Chartered Builder (CIOB) (was *Building Tech. & Management*)
Chartered Quantity Surveyor (now *Building Economist*)
Chartered Surveyor Weekly (now *Property Week*)
CICA Bulletin
CIRIA News
Clerk of Works
Commission of the European Communities
Computer Aided Design
Concrete
Conservation Bulletin
Construction (PSA)
Construction and Building Materials
Construction Engineering Research & Applications (CERA)
Construction Industry Law Letter
Construction Law Digest
Construction Law Journal
Construction Management and Economics
Construction Monitor (replaces *Euronews Construction*)
Construction News
Construction Repair
Construction Safety Information Service
Construction Weekly (now incorporated in *Contract Journal*)
Constructus
Contract Journal (now incorporating *Construction Weekly*)
Croner's Europe
Current Law
Deutschland (German Edition)
Director
Economic and Business Education (previously *Economics*)
Economic Journal
Economics (now *Economic and Business Education*)
Economist
Economy and Society

Energy Management
Engineering Construction and Architectural Management
Environment and Urbanization
Environments by Design: School of Surveying & Architecture, Kingston
Euronews Construction (replaced by *Construction Monitor*)
European Business Journal
European Law Monitor
Financial Management of Property and Construction
Financial Statistics
Forecasts for the UK Economy
Glass Digest (USA)
Habitat I International
Harvey on Industrial Relations and Employment Law
Heating and Ventilating Engineer
House Builder
International Construction Law Review
International Journal of Construction Information Technology
International Journal of Construction Maintenance & Repair
    (now *Construction Repair*)
International Journal of Construction Procurement
International Journal of Project Management
International Journal of Urban & Regional Research
Journal of American Society of Civil Engineers
Journal of Building & Environment: Dept of Architecture, Edinburgh
Journal of Business Law
Journal of Computing in Civil Engineering
Journal of Construction Engineering & Management
Journal of Energy Engineering
Journal of Engineering Mechanics
Journal of Environmental Engineering
Journal of Geotechnical Engineering
Journal of Hydrolics Engineering
Journal of Indoor Air International: Indoor Environment
Journal of Irrigation & Drainage Engineering
Journal of Management in Engineering
Journal of Materials in Civil Engineering
Journal of Microcomputers in Civil Engineering: USA
Journal of Performance of Construction Facilities
Journal of Professional Issues in Engineering
Journal of Property Finance
Journal of Structural Engineering
Journal of Structural Engineering Review

Journal of Surveying Engineering
Journal of Transportation in Engineering
Journal of Urban Planning & Development
Journal of Water Reserves, Planning and Management
Journal of Waterways, Ports, Coastal and Ocean Engineering
Keesings Contemporary Archives
Law Quarterly Review
Le Moniteur
L'Express
Local Economy
Management Today
Managing Service Quality
Master Builder
Materials and Structures
Monthly Bulletin of Construction Indices
New Builder
Northern Economic Review
Oxford Economic Papers
Personal Computer World
Proceedings of Institution of Civil Engineers
Project Manager Today
Property Journal
Property Week (previously *Chartered Survey or Weekly*)
Public Contracts Law Journal: USA
Public Law
Public Procurement Law Review
RICS Abstracts and Reviews
RICS Weekly Briefing
Roofing Cladding and Insulation
Statistical News
Stern
Structural Survey
Surveyor
Taxation
The Structural Engineer
Third World Planning Review
Trade Update
Trade Viewpoint
Valuer
Week In Europe
Weekly Law Reports

Appendix 3
London South Bank University:
Example of a questionnaire

MARKETING QUESTIONNAIRE

Please respond to the following questions by either ticking the appropriate box or by writing your answer in the space provided.

Please note: All information provided will be treated in the strictest of confidence.

Section one – questions related to company details

1.1 Please indicate the approximate turnover of your practice

Less than £2m	£2.1–6m	£6.1–10m	More than £10m (Please State)
☐	☐	☐	_____

1.2 What percentage of this is attributed specifically to marketing (approximately)?

Less than 1%	1–2%	3–5%	More than 5% (Please State)
☐	☐	☐	_____

1.3 Does the practice have a separate marketing department? YES ☐ NO ☐

IF YES

A. How many people are employed is this department? Please tick appropriate box.

1–2	3–5	6–8	More than 8 (Please State)
☐	☐	☐	_____

B. How long has this marketing department been established? Please tick.

0–2 Yrs	3–4 Yrs	5–6 Yrs	More than 6 Yrs (Please State)
☐	☐	☐	_____

IF NO

Please tick one box to indicate for what reasons?

– A marketing department is not considered necessary ☐
– The practice is too small to employ individuals for this ☐
– Marketing tends to be a management role ☐
– Other (Please Specify)

_____

_____

_____

1.4 What does the practice perceive as its future marketing trend over the next 5 years?

– To decrease ☐
– To remain stable ☐
– To increase ☐

1.5 Which statement best describes your organisational philosophy. Please tick one box.

– To achieve modest growth ☐
– To achieve a greater share of the market place ☐
– To achieve service and quality ☐

Section two – questions related to marketing strategy and policy

2.1 What level of priority is given to marketing within your business strategy? Please tick one box.

– High priority ☐
– Moderate priority ☐
– Low priority ☐

2.2 Which statement best describes the role of marketing within your practice? Please tick one box.

- It is perceived to be the role of upper management ☐
- Marketing plays a role with all our staff ☐
- It is left to the marketing department/key individuals ☐

2.3 Which statement best describes the way in which your practice is driven? Please tick one box.

- Technology driven ☐
- Customer driven ☐
- Market driven ☐

2.4 Are planning and operations within departments orientated towards customer satisfaction? Please tick one box.

- Yes very much so ☐
- Yes, in some departments ☐
- No, other factors dominate ☐

2.5 Which statement best describes the way your practice allocates its resources to a project? Please tick one box.

- Resources are governed by the fee and required profit ☐
- Governed by the clients actual requirements ☐
- Governed by the type of client (i.e. New client etc.) ☐

2.6 Is training given to your staff on the basics of marketing? Please tick one or more box(es).

- Yes, we undertake regular training sessions ☐
- We have in the past given staff training ☐
- No, we have never given any formal training ☐

2.7 Does your marketing department seek advice from outside P.R. agencies?

YES ☐   NO ☐

If Yes, For how long? _____ Yrs

If No, for what reasons? (Please Specify) _____

_____

_____

_____

Section three – questions related to the marketing techniques

3.1 From the following marketing and promotional techniques please indicate which techniques does your firm practice by stating (yes/no). If yes, how important do you consider them in making clients aware of your firm.

Marketing/promotional techniques	Do you use Yes/No	← Not important/very important →				
		1	2	3	4	5
Company Brochure						
Company Newsletter						
Seminars						
Presentations						
Videos						
Entertaining						
Advertising through Media						
Other forms of advertising (i.e. site boards etc.)						
Introduction letter						
Company logo						

3.2 If your practice was presenting itself to a client which statement would best describe the initial approach it would use to sell itself? Please tick box(es)

- We would concentrate on past projects, with emphasis on experience and reputation ☐
- We would focus on the particular clients needs, relative to the project ☐
- We would promote the technical facilities and capabilities of the practice ☐
- We would promote value for money and cost ☐

3.3 How often does your practice thoroughly research a new client and the services they require? Please tick box(es)

– We attempt to do this for all clients ☐
– We do research, but not as much as we should ☐
– We tend to offer our own range of services first ☐

3.4 Which statement best describes the primary objectives of your marketing strategy

– To achieve increased market share and new clients ☐
– Service excellence is our primary objective ☐
– Technical excellence and expertise should be exploited ☐

Section four – questions related to the selection process

4.1 In making initial contact with a client, how influential are the following? Please tick appropriate cell.

Forms of Initial Contact	← Least influential/most influential →				
	1	2	3	4	5
Advertisements	___	___	___	___	___
Recommendations	___	___	___	___	___
Introduction Letters/Portfolios	___	___	___	___	___
Industry Reputation	___	___	___	___	___
Previous Working Relationships	___	___	___	___	___
Contacts	___	___	___	___	___
Following Leads	___	___	___	___	___
Others, Please State and Rank:	___	___	___	___	___
_____	___	___	___	___	___
_____	___	___	___	___	___

4.2 How important do you consider the following criteria when being selected by a client for your services? Please tick appropriate cell.

CRITERIA	Little Importance	Some Importance	Quite Important	Important	Very Important
Turnover/Size of practice					
Status/Market share					
Financial standing/ stability					
Technical capability/ excellence					
Quality of personnel and expertise					
Obtaining/having QA					
Being chartered					
Experience					
Professional reputation/ track record					
Prior business relationships/clients					
Lowest price/fees					
Presentation					
Previous projects (type)					
Range of services offered					
Geographical location					
Attention to detail					
Company offices and general image					
Negotiating skill					
Informal contacts					
Responsiveness					

Section five – general questions

5.1 What are the main hurdles/problems in reaching a client? Please give brief details.

_____

_____

_____

5.2 Do you feel there is any way QS Practices can improve the services they offer (if necessary)?

_____

_____

_____

_____

Thank you for your participation, please send the completed questionnaires in the self-addressed envelope provided.

Appendix 4
Statistical tables

Table A *Distribution of* t

DF	Probability					
	0.5	0.1	0.05	0.02	0.01	0.001
1	1.000	6.314	12.706	31.821	63.657	636.619
2	0.816	2.920	4.303	6.965	9.925	31.598
3	0.765	2.353	3.182	4.541	5.841	12.941
4	0.741	2.132	2.776	3.747	4.604	8.610
5	0.727	2.015	2.571	3.365	4.032	6.859
6	0.718	1.943	2.447	3.143	3.707	5.959
7	0.711	1.895	2.365	2.998	3.499	5.405
8	0.706	1.860	2.306	2.896	3.355	5.041
9	0.703	1.833	2.262	2.821	3.250	4.781
10	0.700	1.812	2.228	2.764	3.169	4.587
11	0.697	1.796	2.201	2.718	3.106	4.437
12	0.695	1.782	2.179	2.681	3.055	4.318
13	0.694	1.771	2.160	2.650	3.012	4.221
14	0.692	1.761	2.145	2.624	2.977	4.140
15	0.691	1.753	2.131	2.602	2.947	4.073
16	0.690	1.746	2.120	2.583	2.921	4.015
17	0.689	1.740	2.110	2.567	2.898	3.965
18	0.688	1.734	2.101	2.552	2.878	3.922
19	0.688	1.729	2.093	2.539	2.861	3.883
20	0.687	1.725	2.086	2.528	2.845	3.850
21	0.686	1.721	2.080	2.518	2.831	3.819
22	0.686	1.717	2.074	2.508	2.819	3.792
23	0.685	1.714	2.069	2.500	2.807	3.767
24	0.685	1.711	2.064	2.492	2.797	3.745
25	0.684	1.708	2.060	2.485	2.787	3.725
26	0.684	1.706	2.056	2.479	2.779	3.707
27	0.684	1.703	2.052	2.473	2.771	3.690
28	0.683	1.701	2.048	2.467	2.763	3.674
29	0.683	1.699	2.045	2.462	2.756	3.659
30	0.683	1.697	2.042	2.457	2.750	3.646
40	0.681	1.684	2.021	2.423	2.704	3.551
60	0.679	1.671	2.000	2.390	2.660	3.460
120	0.677	1.658	1.980	2.358	2.617	3.373
∞	0.674	1.645	1.960	2.326	2.576	3.291

Source: Adapted by permission of the authors and publishers from Table III of Fisher and Yates, *Statistical Tables for Biological, Agricultural and Medical Research*, published by Longman Group Ltd., London (previously published by Oliver and Boyd, Edinburgh).

Table B *Distribution of* χ^2

DF	0.5	0.1	0.05	0.02	0.01	0.001
			Probability			
1	0.455	2.706	3.841	5.412	6.635	10.827
2	1.386	4.605	5.991	7.824	9.210	13.815
3	2.366	6.251	7.815	9.837	11.345	16.268
4	3.357	7.779	9.488	11.668	13.277	18.465
5	4.351	9.236	11.070	13.388	15.086	20.517
6	5.348	10.645	12.592	15.033	16.812	22.457
7	6.346	12.017	14.067	16.622	18.475	24.322
8	7.344	13.362	15.507	18.168	20.090	26.125
9	8.343	14.684	16.919	19.679	21.666	28.877
10	9.342	15.987	18.307	21.161	23.209	29.588
11	10.341	17.275	19.675	22.618	24.725	31.264
12	11.340	18.549	21.026	24.054	26.217	32.909
13	12.340	19.812	22.362	25.472	27.688	34.528
14	13.339	21.064	23.685	26.873	29.141	36.123
15	14.339	22.307	24.996	28.259	30.578	37.697
16	15.338	23.542	26.296	29.633	32.000	39.252
17	16.338	24.769	27.587	30.995	33.409	40.790
18	17.338	25.989	28.869	32.346	34.805	42.321
19	18.338	27.204	30.144	33.867	36.191	43.820
20	19.337	28.412	31.410	35.020	37.566	45.315
21	20.337	29.615	32.671	36.343	38.932	46.797
22	21.337	30.813	33.924	37.659	40.289	42.268
23	22.337	32.007	35.172	38.968	41.638	49.728
24	23.337	33.196	36.415	40.270	42.980	51.179
25	24.337	34.382	37.652	41.566	44.314	56.620
26	25.336	35.563	38.885	42.856	45.642	54.052
27	26.336	36.741	40.113	44.140	46.963	55.476
28	27.336	37.916	41.337	15.419	48.278	56.893
29	28.336	39.087	42.557	46.693	49.588	58.302
30	29.336	40.256	43.773	47.962	50.892	59.703

Source: Adapted by permission of the authors and publishers from Table IV of Fisher and Yates, *Statistical Tables for Biological, Agricultural and Medical Research*, published by Longman Group Ltd., London (previously published by Oliver and Boyd, Edinburgh).

Table C   *Critical values of ρ (rho) at various levels of probability (Spearman rank correlation coefficient)*
For any N the observed value of ρ, is significant at a given level of significance if it is *equal* to or *larger* than the critical values shown in the table

N (number of subjects)	Level of significance for one-tailed test			
	.05	.025	.01	.005
	Level of significance for two-tailed test			
	.10	.05	.02	.01
5	.900	1.000	1.000	–
6	.829	.886	.943	1.000
7	.714	.786	.893	.929
8	.643	.738	.833	.881
9	.600	.683	.783	.833
10	.564	.648	.746	.794
12	.506	.591	.712	.777
14	.456	.544	.645	.715
16	.425	.506	.601	.665
18	.399	.475	.564	.625
20	.377	.450	.534	.591
22	.359	.428	.508	.562
24	.343	.409	.485	.537
26	.329	.392	.465	.515
28	.317	.377	.448	.496
30	.306	.364	.432	.478

Note: When there is no exact number of subjects use the next lowest number.

Table D *Critical values of r at various levels of probability (Pearson product movement correlation)*

For any particular N the observed value of r is significant at a given level of significance if it is equal to or larger than the critical values shown in the table

	Level of significance for one-tailed test				
	.05	.025	.01	.005	.005
	Level of significance for two-tailed test				
df = N − 2	.10	.05	.02	.01	.001
1	.9877	.9969	.9995	.9999	1.0000
2	.9000	.9500	.9800	.9900	.9990
3	.8054	.8783	.9343	.9587	.9912
4	.7293	.8114	.8822	.9172	.9741
5	.6694	.7545	.8329	.8745	.9507
6	.6215	.7067	.7887	.8343	.9249
7	.5822	.6664	.7498	.7977	.8982
8	.5494	.6319	.7155	.7646	.8721
9	.5214	.6021	.6851	.7348	.8471
10	.4793	.5760	.6581	.7079	.8233
11	.4762	.5529	.6339	.6835	.8010
12	.4575	.5324	.6120	.6614	.7800
13	.4409	.5139	.5923	.6411	.7603
14	.4259	.4973	.5742	.6226	.7420
15	.4124	.4821	.5577	.6055	.7246
16	.4000	.4683	.5425	.5897	.7084
17	.3887	.4555	.5285	.5751	.6932
18	.3783	.4438	.5155	.5614	.6787
19	.3687	.4329	.5034	.5487	.6652
20	.3598	.4227	.4921	.5368	.6524
25	.3233	.3809	.4451	.4869	.5974
30	.2960	.3494	.4093	.4487	.5541
35	.2746	.3246	.3810	.4182	.5189
40	.2573	.3044	.3578	.3932	.4896
45	.2428	.2875	.3384	.3721	.4648
50	.2306	.2732	.3218	.3541	.4433
60	.2108	.2500	.2948	.3248	.4078
70	.1954	.2319	.2737	.3017	.3799
80	.1829	.2172	.2565	.2830	.3568
90	.1726	.2050	.2422	.2673	.3375
100	.1638	.1946	.2301	.2540	.3211

Note: When there is no exact df use the next lowest number.

Appendix 5
Construction industry employee injury statistics

1994/1995 (provisional) Injuries to employees in the construction industry, as reported to HSE's Field Operations Division

Occupation	Fatal	Major	Over 3 days	Total
Bricklayer	2	72	550	624
Carpenter/joiner	2	136	1127	1265
Demolition worker	–	16	37	53
Electrician	–	108	330	438
Glazier	1	12	96	109
Ground worker	4	58	139	201
Painter, decorator	–	61	233	294
Pavior, roadman	3	39	443	485
Pilling hand	1	6	33	40
Plasterer	–	22	158	180
Plumber, pipe fitter	–	71	543	614
Scaffolder	2	62	222	286
Slater, roof worker	7	64	195	266
Steel erector	–	24	36	60
Steel fixer, bar bend	–	10	30	40
Steeplejack	–	6	2	8
Welder	–	20	72	92
Other const. trade	5	109	378	492
Drivers	4	120	539	663
Electrical linesman	–	8	23	31
Electrical fitter	–	10	50	60
Electrical jointer	–	4	14	18
Communicat. Eng.	–	1	8	9
Labourer	9	316	1659	1984
Mainten. personnel	5	60	294	359
All managerial	2	120	473	595
Manual production	2	104	548	654
Other occupations	–	78	623	701
Not known	–	81	520	601
TOTAL	49	1798	9375	11222

Index